A PRIMER OF
DIFFUSION PROBLEMS

Reprinted by permission from Chas Addams.

A PRIMER OF DIFFUSION PROBLEMS

Richard Ghez
IBM Thomas J. Watson Research Center
Yorktown Heights, New York

WILEY

A Wiley-Interscience Publication

John Wiley & Sons

New York • Chichester • Brisbane • Toronto • Singapore

This book was set in press type by the author on an Autologic APS5 printer driven by an IBM VM/3081-K system. The formatting was performed using Termtext/Format and the Yorktown Mathematical Formula Processor.

Library of Congress Cataloging in Publication Data:

Ghez, Richard.
   A primer of diffusion problems.

   Includes bibliographies and index.
   1. Diffusion processes—Mathematical models.
2. Differential equations, Partial. I. Title.
QA274.75.G48   1988   515.3'53   87-37284
ISBN 0-471-84692-9

Printed in the United States of America

10 9 8 7 6 5 4 3 2 1

A ma mère et a ma femme
qui, aussi, savent croquer la pomme.

PREFACE

Most kinetic processes in ponderable media require the redistribution of mass, energy, momentum, or charge over macroscopic distances. This redistribution occurs through a random process called diffusion, and its macroscopic description rests on a particular differential equation called the diffusion equation. Because this equation is to ponderable media what the wave equation is to vacuum, it is no wonder that its solutions should be relevant to widely different problems: from the heat treatment of alloys to the motion of proteins in growing axon membranes; from the dynamics of galactic structures to the multitude of device-processing steps. Consequently, the literature on diffusion contains at least four different points of view: as a mathematical topic in partial differential equations, as a physico-chemical topic in non-equilibrium thermodynamics, as an engineering topic in materials science, and as a topic in stochastic processes. It would be presumptuous to claim familiarity with all these topics and to suggest, here, an equal coverage. And yet, they all derive from the same basic principle, namely, that something or other is *conserved* or can be *balanced*. Conservation is not so much an expression of longevity — as in "this person is well preserved" — but rather it expresses our ability to distinguish and to reckon. Nowhere, to my knowledge, is this point of view better illustrated than in the frontispiece, and I am deeply indebted to Mr. Chas Addams for his permission to reproduce an example of his always penetrating wit.

In spite of this formal unity of diffusion problems, it is my experience that many scientists and engineers remain baffled by the diffusion equation and by its solution in practical cases. Alternatively, they are perhaps trapped between purely mathematical texts and others that deal excusively with diffusion mechanisms. This book seeks to bridge the gap between physico-chemical statements of certain kinetic processes and their reduction to diffusion problems. It also attempts to introduce the reader to the many lines of attack, both analytic and numerical, on the diffusion equation. I have chosen to teach through physically significant examples taken mainly from my experience in the areas of metallurgy and of semiconductor technology. The theory is interwoven, I hope, with sufficient rigor so that the tools that evolve do not immediately crumble with use. Nevertheless, I have chosen to write an introductory text, whose only prerequisites are a serious year of calculus, through ordinary differential equations, one semester of thermodynamics, and, if the reader be uneasy with algebraic manipulations, that the fear of calculating be controllable.

As its title implies, this is a also book about *solving* the diffusion equation. I also wanted a short and lively text. Consequently, unless they provide particular insight, intermediate steps of specific calculations are often only verbally (though completely) expressed. In other words, the beginner would be well advised to arm himself with pad and pencil. Exercises, not so numerous as to paralyze the reader, are sprinkled throughout the text. They are designed either to emphasize a point that has just been made or to suggest some further clarification or extension of that point. These exercises are an integral part of the text, and (a word of encouragement) none is so difficult as to constitute a doctoral thesis.

The text consists of seven chapters and three appendices. These appendices contain more advanced material, whose inclusion within chapters would have detracted from the general flow. References, cited in the text in square brackets, are listed at the end of each chapter, and, in general, equations are numbered decimally within each chapter. I have labeled equations, however, with an additional Latin letter when these are closely related. It can happen, therefore, that "well ordering" be not everywhere preserved.

I might be faulted on at least one count. My choice of topics is definitely biased, and each specialist will surely be disappointed to find that his particular area of expertise was not covered. This is unavoidable in so short a text, and I will simply observe that presentations of methodology and techniques are useful even though they fail as encyclopedias.

If any text is a slice of the author's life, it is also an image of his relations with others from whom he has benefited. I am grateful to my employer, the IBM Corporation, for allowing me the time and the use of its facilities to complete this book. Moreover, my experience with diffusion would be insignificant were it not for discussions and collaboration with many colleagues in this and other laboratories. Among these, I wish to single out my friends E.A. Giess, M.B. Small, F.M. d'Heurle, T.-S. Kuan, all at IBM, and A.S. Jordan and G.H. Gilmer at AT&T, Bell Laboratories. J.S. Lew and L. Kristianson, also at IBM, are gratefully thanked for their kindness and patience with my many queries. They are holders of increasingly rare skills (mathematical analysis and technical illustration) in a world that too often confuses quality with touch typing. I am also deeply indebted to Bea Shube and to her staff at John Wiley & Sons for their encouragement and help, particularly during those dark periods that every author knows. Finally, I wish to recall the memory of M.M. Faktor (Queen Mary College, London), a true and courageous friend and scientist.

R. Ghez

Yorktown Heights, NY
February, 1988

CONTENTS

Chapter 6. A User's Guide to the Laplace Transform   167

Chapter 7. Further Time-Dependent Examples   193

Appendices   223

Index   241

A PRIMER OF
DIFFUSION PROBLEMS

Chapter 1

THE DIFFUSION EQUATION

We begin with a model for diffusion: the isotropic one-dimensional random walk [1-4]. It is so simple that the basic physical processes cannot elude us. It also has a continuum limit, the diffusion equation, whose solutions are here our main concern, and some of whose properties we then examine. Conversely, this model forms the basis for numerical methods of solution. We then discuss the diffusion equation's form in higher dimensions and other physical instances where that equation offers a realistic description. This chapter ends with a brief account of the origin of conservation principles and constitutive relations that pervade transport phenomena.

1.1 ISOTROPIC ONE-DIMENSIONAL RANDOM WALK

Consider points on a line, as shown in Fig. 1.1, choose an arbitrary origin, and label these points through integers $i = 0, \pm 1, \pm 2, \dots$. Then attribute particles to each point, henceforth called sites, i.e., N_i is the number of particles at site i. Let us assume that each of these particles can jump to adjacent sites with a frequency Γ that does not depend on the site i. Jumps to the right are, on average, as likely as those to the left, so that $\frac{1}{2}\Gamma N_i$ particles jump from site i to site $i+1$ per unit time.†
The reverse transition $i+1 \rightarrow i$ occurs at the rate $\frac{1}{2}\Gamma N_{i+1}$, and it follows

---

† Contrary to certain claims, it is not $N_i/2$ particles that transit, on average, in either direction, but rather N_i particles that have a transition frequency (or transition probability per unit time) $\Gamma/2$ of carrying out a specific jump. Thus, Γ must be understood as the *total* transition probability. It is also a *conditional* probability per unit time in the sense that a transition occurs, *given* that an initial site is occupied [cf. remark (d) below].

1

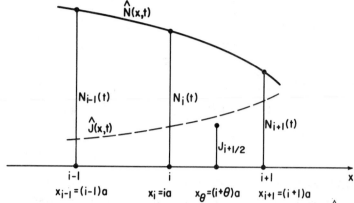

Fig. 1.1: Particle numbers N_i at sites i, the continuous interpolation $\hat{N}$, and the flux $\hat{J}$ derived therefrom. Note the "gap" between $\hat{J}$ and $J_{i+1/2}$, which is minimum at the midpoint.

that the *net flux* (or rate of exchange) of particles that hop from site i to site $i+1$ per unit time is

$$J_{i+1/2} = \frac{1}{2}\Gamma(N_i - N_{i+1}).$$ (1.1a)

Likewise, the net flux from $i-1$ to i is

$$J_{i-1/2} = \frac{1}{2}\Gamma(N_{i-1} - N_i).$$ (1.1b)

The subscript notation $i \pm 1/2$, admittedly somewhat obscure (the N's are defined only at integer sites), will later become clear. For the moment, it simply indicates that the particle fluxes (1.1) depend on *both* N_i and N_{i+1}. These fluxes vanish when $N_i = N_{i+1}$ for all i; in other words, the flow of matter vanishes when mass is evenly distributed on the sites i. This is nothing but a necessary condition for equilibrium.

Fluxes are fine, but we are really more interested in the particle distribution N over the sites i. Towards that end, compute its time rate of change by evaluating all possible transitions to and from the ith site:

$$\frac{dN_i}{dt} = -\frac{1}{2}\Gamma N_i + \frac{1}{2}\Gamma N_{i+1} - \frac{1}{2}\Gamma N_i + \frac{1}{2}\Gamma N_{i-1}.$$

Collecting terms, we get a rate equation in terms of the particle distribution

$$\dot{N}_i = \tfrac{1}{2}\Gamma(N_{i+1} + N_{i-1} - 2N_i), \tag{1.2}$$

or, using Eqs. (1.1), in terms of fluxes

$$\dot{N}_i = -(J_{i+1/2} - J_{i-1/2}). \tag{1.3}$$

Equation (1.2) gives the rate of change in terms of the nearest-neighbor distribution around site i, whereas Eq. (1.3) expresses that change as a difference of fluxes. In a sense, this last expression is the more fundamental, for it is a statement of a *conservation law*. Imagine a cell centered at site i, whose boundaries are somewhere in the intervals $[i-1,i]$ and $[i,i+1]$. Then Eq. (1.3) simply states that the change in the cell's content is solely due to the difference between the "stuff in" $J_{i-1/2}$ and the "stuff out" $J_{i+1/2}$. In other words, this equation is nothing but a statement of (local) mass conservation, and it would still be true even if the fluxes had forms different from (1.1), *e.g.*, polynomial functions of the differences N_i-N_{i+1}, or functions involving next nearest-neighbor jumps. Let us pause for a few

Remarks:
   a. Equation (1.2) is linear in the unknown functions $N_i(t)$. Structurally, it is a linear difference-differential equation, second-order in i and first-order in t. Given an initial configuration (initial condition) $N_i(0) = N_i^{(0)}$ and boundary conditions (on i), this initial-value problem can be solved analytically. We will not develop this here. However, Eq. (1.2) does provide a numerical scheme for its continuum analog, as Sect. 1.4 will show.
   b. The particular simplicity of our random walk rests on the constancy of the jump frequency Γ. Recall that Γ is assumed independent of the site index i and of time. In addition, Γ does not depend on the actual particle distribution in the neighborhood of an ith site. Succinctly, the hopping model is *homogeneous*. This assumption must often be relaxed, for example in vacancy diffusion: A jump can occur only if adjacent sites are vacant. As pointed out in the first footnote, the model is also *isotropic* because jumps to the right and to the left are

equally likely: $\frac{1}{2}\Gamma$ represents either frequency if Γ is the total jump freqency. Later, in Chap. 3, we shall see that external forces introduce a measure of anisotropy.

c. Equations (1.1-3) are topological in the sense that the *distance* between points or the *shape* of the lattice play no role whatsoever. Only the *ordering* of points matters: One must always be able to tell which of two adjacent points is "the first." For example, these equations are valid for points distributed along a space curve — a helix, say — but points of self-intersection, as in a "figure eight," require special attention. In brief, these equations are valid for spaces topologically equivalent to the real line.

d. The reader will note the similarity between Eqs. (1.1-3), applied here to mass transport, and some of those used to describe nuclear decay, chemical kinetics, and random processes. In fact, similar equations describe the time evolution of any property over an ordered list of *states i.* If $P_i(t)$ is that property, and $\Gamma_{i,k}$ is the transition frequency from state i to k, then

$$\dot{P}_i = \sum_k (\Gamma_{k,i} P_k - \Gamma_{i,k} P_i) \tag{1.4}$$

is a generalization of Eq. (1.2), often called a "master equation." Its probabilistic interpretation is interesting: If P_i is the *a priori* occupation probability of state i and $\Gamma_{i,k}$ is the *conditional* probability of the $i \rightarrow k$ transition (given that i is occupied), then sums such as $\Sigma_i \Gamma_{i,k} P_i$ represent the *total* probability of transition to a given final state k, regardless of the initial state i. Both sums of Eq. (1.4) can be interpreted accordingly.

1.2 CONTINUUM EQUATIONS

In many instances we may be unconcerned with the discrete nature of matter. It is then possible to find a continuum analog of Eqs. (1.1-3). The essential step requires that the sites i be points of the real line, *i.e.*, we must introduce a distance, or "metric," between points. Assume, then, that the sites are equidistant, a being the lattice (or jump) distance. Therefore, the ith site has the coordinate $x_i = ia$. Now introduce *any* continuous and sufficiently differentiable function $\hat{N}(x,t)$ that interpolates the previous functions $N_i(t)$, namely, that satisfies

$$\hat{N}(x_i,t) = N_i(t) \tag{1.5}$$

at sites $x = x_i$, but that is quite arbitrary elsewhere. This is shown schematically in Fig. 1.1. Omitting the time variable for the moment, expand the interpolating function in a Taylor series around x_i. For adjacent points, we get†

$$\hat{N}(x_{i\pm1}) = \hat{N}(x_i) \pm a\frac{\partial \hat{N}}{\partial x}\Big|_{x_i} + \frac{1}{2}a^2\frac{\partial^2 \hat{N}}{\partial x^2}\Big|_{x_i} + O(a^3), \tag{1.6a}$$

which, with the condition (1.5), becomes

$$N_{i\pm1} = N_i \pm a\frac{\partial \hat{N}}{\partial x}\Big|_{x_i} + \frac{1}{2}a^2\frac{\partial^2 \hat{N}}{\partial x^2}\Big|_{x_i} + O(a^3). \tag{1.6b}$$

Introducing this formula into Eq. (1.2), we easily get

$$\frac{\partial \hat{N}}{\partial t} = \frac{1}{2}\Gamma a^2\frac{\partial^2 \hat{N}}{\partial x^2} + O(a^4). \tag{1.7}$$

This is a *single* partial differential equation for the function $\hat{N}$, which replaces the *denumerable* (possibly infinite) coupled system of difference-differential equations (1.2).‡

---

† The "O symbol" in these and other equations has the usual mathematical meaning: $f(a) = O[g(a)]$ if the ratio f/g tends to a finite limit as $a \to 0$. It can be read as "plus terms of order...."

‡ How good an approximation is Eq. (1.7) to the discrete form (1.2)? The question is easily answered for a particular solution: the plane wave $\hat{N}(x,t) = A_k(t) \sin kx$. Any solution will later be seen to be a linear superposition of such plane waves with wavelengths $\lambda = 2\pi/k$, so that there is no loss of generality. If we systematically carry out the expansions (1.6) to higher order, then it is easily seen that all terms in Eq. (1.7) of odd order in a vanish identically and that the first truncated term is $(\Gamma a^4/24) \partial^4 \hat{N}/\partial x^4$. Introducing the plane wave solution into Eq. (1.7), to which we add this fourth-order term, we get an equation $\dot{A}_k/A_k = -(1/2)\Gamma a^2k^2(1-a^2k^2/12)$ for the amplitudes. The truncated expression (1.7) is valid if $ak \ll 1$, or, in words: the jump distance must be smaller than the smallest significant wavelength. This means that the amplitudes A_k must be insignificant outside of a "cutoff" $|k_{max}|$.

It is somewhat more delicate to analyze the fluxes, for it is *a priori* not at all evident how to define a continuous analog $J(x,t)$ for the flux. Again introducing formula (1.6b) into Eqs. (1.1), we find that

$$J_{i\pm1/2} = -\frac{1}{2}\Gamma\left(a\frac{\partial\hat{N}}{\partial x} \pm \frac{1}{2}a^2\frac{\partial^2\hat{N}}{\partial x^2} \right)\Big|_{x_i} + O(a^3). \qquad (1.8)$$

In other words, both fluxes converge to the same value $-\frac{1}{2}\Gamma a\,\partial\hat{N}/\partial x$ (evaluated at x_i), to first order in a, and the second order terms are equal and opposite. It follows that Eqs. (1.8), inserted into Eq. (1.3), again yield the expression (1.7), but that we have no continuous analog of that discrete conservation law (1.3) unless we retain second-order terms in Eq. (1.8). Therefore, let us agree to *define for all x* a continuous flux function

$$\hat{J}(x,t) = -\frac{1}{2}\Gamma a\frac{\partial\hat{N}}{\partial x} \qquad (1.9)$$

in terms of the interpolating function $\hat{N}$. In other words, a continuous flux function can be calculated if $\hat{N}$ is known, but it is quite unlikely that $\hat{J}$ will ever pass exactly through the discrete fluxes (1.1), no matter where these are placed. The "gap" can be minimized, however, if we measure the difference between $J_{i+1/2}$, say, and expression (1.9) evaluated at arbitrary points in the interval $[i,i+1]$. Specifically, define an arbitrary intermediate point $x_\theta = (i+\theta)a$, where $0 \le \theta \le 1$. Form, then, the difference $J_{i+1/2} - \hat{J}(x_\theta)$, and expand all quantities around x_θ. The result is given in the form of an

Exercise 1.1: *To justify the notation $J_{i\pm1/2}$*
Show that the above difference in fluxes is minimum when $\theta = \frac{1}{2}$. What else can you say about the series in θ? Then develop the same arguments for the flux $J_{i-1/2}$ in the interval $[i-1,i]$, and show that it has the *same* formal expansion around the appropriate intermediate point as does $J_{i+1/2}$. Finally, form the difference $J_{i+1/2} - J_{i-1/2}$, expand it around the current lattice site x_i, and show that it is equal to $a(\partial\hat{J}/\partial x)|_{x_i}$ plus fourth-order terms in a. ∎

It follows from this exercise that the discrete conservation law (1.3) has a continuum analog:

$$\frac{\partial \hat{N}}{\partial t} = -a\frac{\partial \hat{J}}{\partial x} + O(a^4).$$

(1.10)

We have now reached our goal, for, if $C = \hat{N}/a$ is the average concentration per cell, and if we introduce the definition

$$D = \frac{1}{2}\Gamma a^2$$

(1.11)

of the *diffusion coefficient* or *diffusivity*, then Eqs. (1.9), (1.10), and (1.7) become

$$J(x,t) = -D\frac{\partial C}{\partial x},$$

(1.12)

$$\frac{\partial C}{\partial t} = -\frac{\partial J}{\partial x},$$

(1.13)

$$\frac{\partial C}{\partial t} = D\frac{\partial^2 C}{\partial x^2}.$$

(1.14)

Exercise 1.2: *Next-nearest-neighbor jumps*
Let Γ_1 and Γ_2 be the total jump frequencies to nearest and next nearest neighbors, respectively. Show that Eq. (1.14) is still valid if the diffusivity is defined appropriately. Can you generalize? This example is of some interest because it shows that two distinct difference equations can yield the same continuum equation. ▌

Remarks:
    a. Equations (1.12) and (1.14) are sometimes called Fick's first and second laws [3]. This is but nomenclature. It *is* important, however, to remember that Eq. (1.13), the so-called continuity equation, is a precise statement of mass conservation, as we shall soon see. On the other hand, the relation (1.12) between flux J and thermodynamic force $\partial C/\partial x$, sometimes called a *constitutive relation*, does not have the same generality. Its form depends on the details of the diffusion mechanism. Note, nonetheless, that its minus sign means that material must flow from regions of high concentration to regions of lower concentration, a property that we ordinarily associate with diffusion. We will

refer, henceforth, to Eq. (1.14) as the *diffusion equation*; it is a consequence of the two previous equations if the diffusivity is a constant.

b. The diffusion equation is sometimes also called the heat equation [5,6], because the evolution of energy also obeys a conservation law. From a mathematical point of view, it is the simplest so-called *parabolic* equation (more nomenclature). From the theory of partial differential equations [7] we learn that it must be solved as an initial-value problem: Given initial data on C and boundary conditions on the spatial variable x, we can then find the solution by "marching forward" in time. This will soon become clear in Sects. 1.4-5 on numerical methods.

c. From Eq. (1.11) we learn that the constancy of D is related to the constancy of the jump distance and frequency. If, for any reason, these are not constants (see the next exercise), then Eq. (1.14) must be replaced by the more general

$$\frac{\partial C}{\partial t} = \frac{\partial}{\partial x}\left(D \frac{\partial C}{\partial x} \right). \tag{1.15}$$

d. Physical dimensions are important, particularly if we wish to cast a given problem in as few variables as possible. We shall always denote dimensions by square brackets. Thus from Eq. (1.11) we find that $[D] = \text{cm}^2/\text{s}$. The dimensions of C and J depend on that choice for N_i or $\hat{N}$. For example, if the particles are evenly distributed along planes perpendicular to the direction of diffusion, the usual case of "volume diffusion," then we have $[N_i] = \text{cm}^{-2}$, $[C] = \text{cm}^{-3}$, and $[J] = \text{cm}^{-2}\,\text{s}^{-1}$. Can you find these dimensions if the particles are evenly distributed along lines (surface diffusion) or restricted only to sites (edge diffusion)? Note also that mass or molar units, rather than numbers of particles, can be used to measure N_i, with concomitant changes for C and J.

e. Order-of-magnitude estimates of D are instructive. If, for Γ, we take the Debye frequency in a solid $\sim 10^{13}\ \text{s}^{-1}$, and for the jump distance a we take $\sim 2\ \overset{\circ}{\text{A}}$, then Eq. (1.11) tells us that $D \cong 2 \times 10^{-3}\ \text{cm}^2/\text{s}$. This is many orders of magnitude larger than actually measured in solids. The moral is simply that not all atomic oscillations lead to successful jumps: Only big ones succeed, as is true in other random walks of life.

f. The careful reader will have noticed that the "hat" has come off the symbol J for the flux somewhere between Eqs. (1.9) and (1.12). This is not an oversight. There are two points to bear in mind regarding fluxes: First, they are directed quantities; the function J must then be the component of a flux vector along the jump direction. Second, by a similar construction in more than one dimension it becomes clear that the fluxes, as defined here, are quantities integrated over lines or surfaces transverse to the jump direction. The true flux derives from $\hat{J}$ by dividing by appropriate lengths or areas. This remark is related to remark (d) above; a fuller account is provided in Sect. 1.6 and in Appendix A.

Exercise 1.3: *A model for variable diffusivity*
Write first the master equation (1.4) restricted to nearest-neighbor jumps. Then assume that the transition frequencies depend only on the occupation of the final states: $\Gamma_{i,i\pm 1} = \frac{1}{2}\Gamma f_{i\pm 1}$ for all i, where the dimensionless and arbitrary function $f_{i\pm 1}$ depends only on the occupation numbers $N_{i\pm 1}$ of adjacent states. Show that Eq. (1.15) follows from Eq. (1.4), with a concentration-dependent diffusivity $D(C) = \frac{1}{2}\Gamma a^2 f[1-(d\ln f)/(d\ln C)]$. Can you interpret this? (*cf.* Ref. [8].) ∎

1.3 ELEMENTARY PROPERTIES OF THE DIFFUSION EQUATION

The diffusion equation enjoys several interesting properties [5-7] that are useful, either to characterize the behavior of the solutions, or to derive new solutions from known ones. These properties can be subdivided into three categories: Conservation, smoothing, and invariance.

A.1 Conservation in continuous regions
The conservative nature of Eqs. (1.13) or (1.14-15), or of their discrete analogs (1.2-3), has already been emphasized. We are now ready for a disarmingly simple proof. Consider a diffusion process in a stationary "slab," and denote by x_L and x_R the coordinates of its left- and right-hand boundaries, respectively. The total mass contained in this region is then

$$M(t) = \int_{x_L}^{x_R} C(x,t)\,dx. \tag{1.16}†$$

Taking its time derivative and using Eq. (1.13), we get

$$\dot{M} = \int_{x_L}^{x_R} \frac{\partial C}{\partial t}\,dx = -\int_{x_L}^{x_R} \frac{\partial J}{\partial x}\,dx = J(x_L,t) - J(x_R,t). \tag{1.17}$$

This string of equalities can be viewed in two ways: Either the container (the slab's boundary) is impermeable, in which case the two boundary fluxes in the above expression are zero, and the total mass M is constant, or the constancy of M implies, at the very least, that whatever enters the container must also leave. This characterizes the "steady state," to which we shall return in the next chapter. The very same operations can be performed on the discrete form (1.3). Note that they are independent of the constitutive relations (1.1) or (1.12). At the end of this chapter we shall see a converse statement, namely, global mass conservation in continuous regions implies the continuity equation (1.13).

A.2 Conservation at fixed surfaces of discontinuity

The previous calculations assumed that there were no points of discontinuity within the slab. What if the slab is made of two distinct materials or of two distinct phases, each characterized by its own diffusivity? If $x = \xi$ is such a point separating the two materials or phases, then it is only required that we break up the integral (1.16) into two parts:

$$M = \int_{x_L}^{\xi} C_L\,dx + \int_{\xi}^{x_R} C_R\,dx. \tag{1.18}$$

For simplicity, assume that the fluxes at the boundaries, $J_L(x_L,t)$ and $J_R(x_R,t)$, either vanish or are equal and that the total mass is constant. Then manipulations similar to those leading to Eq. (1.17) yield

---

† The word "mass" is used loosely, for the integral (1.16) over concentration depends on the units of C.

$$\dot{M} = 0 = J_R(\xi+,t) - J_L(\xi-,t). \tag{1.19a}†$$

In other words, the flux is always continuous at stationary internal boundaries, as long as these boundaries are not the seat of other processes that change the number of particles (see a footnote below). The concentration distribution, however, is not smooth. With Eq. (1.12) we have

$$D_L \frac{\partial C_L}{\partial x}\Big|_{\xi-} = D_R \frac{\partial C_R}{\partial x}\Big|_{\xi+}, \tag{1.19b}$$

which shows that $C(x)$ suffers a change in slope if the diffusivities are unequal.

A.3 Conservation at a moving boundary

Not all boundaries are fixed; in fact some of the more interesting ones execute motions, as we shall see in the examples. It is then required, in the last calculations, to relax only one assumption: The internal boundary point's position is now an arbitrary function of time, $x = \xi(t)$. The integrals (1.18) then have variable limits, and, using Leibnitz's rule‡ for such cases, we easily get a generalization of Eqs. (1.19):

$$\dot{\xi}\left(C_L\Big|_{\xi-} - C_R\Big|_{\xi+}\right) = J_L\Big|_{\xi-} - J_R\Big|_{\xi+}$$

$$= -D_L \frac{\partial C_L}{\partial x}\Big|_{\xi-} + D_R \frac{\partial C_R}{\partial x}\Big|_{\xi+} \tag{1.20}$$

In other words, the quantity $\dot{\xi}C - J$ is now continuous across a moving boundary, but the jump in concentration and flux at that boundary are related through its velocity $\dot{\xi}$. Concentration fields often *do* suffer discontinuities at phase boundaries, although, in general, the tempera-

---

† The symbols $\xi\pm$ have their usual mathematical meaning as right and left limits at the point of discontinuity $x = \xi$.

‡ This rule of calculus reads

$$(d/dt)\int_{a(t)}^{b(t)} f(x,t)\,dx = \int_{a(t)}^{b(t)} (\partial/\partial t)f(x,t)\,dx + \dot{b}f(b,t) - \dot{a}f(a,t).$$

ture field for analogous melting and freezing problems does not.† In that case, one can show [5,7] that the left-hand side of Eq. (1.20) is replaced by $\dot{\xi}L$, where L is the latent heat per unit volume. Relations such as Eq. (1.20) are called *Stefan conditions*; they are, in fact, equations of motion for moving boundaries. Any failure to satisfy Stefan conditions is equivalent to a violation of mass or energy conservation.‡

Exercise 1.4: *Geometric interpretation of flux continuity*
Show that the flux continuity condition (1.19b) for a stationary boundary leads to a "law of refraction." Discuss the possible signs of the gradients at the interface. How is this modified if the boundary can move according to Eq. (1.20)? ∎

B. Smoothing property

The diffusion equation (1.14) has an interesting geometric interpretation if we think of the concentration distribution as a surface $z = C(x,t)$ over the plane of the independent variables (x,t). The time derivative $\partial C/\partial t$ is proportional to the local curvature $\partial^2 C/\partial x^2$ along isochronal lines (*i.e.*, a *profile*) of the surface. Figure 1.2 shows a "snapshot" of that surface. Now, because the diffusivity is positive, bumps with negative curvature will tend to move down in time, and positive curvature bumps will move up. In other words, a bumpy profile will become smoother as time progresses. Figure 1.3 shows a particular concentration surface that we will discuss at length in Chap. 4. The vectors $(0, 1, \partial C/\partial t)$, tangent to lines of constant x and directed towards increasing values of t, indeed point downwards where the curvature is negative. Can you make similar statements for Eq. (1.15)?

---

† This can be understood if, for example, processes at the moving boundary occur fast enough for *local* equilibrium to obtain. Equilibrium then implies the continuity of the chemical potential and of the temperature field across that boundary (see Sect. 2.4). But equal chemical potentials μ generally imply a discontinuity in concentration, because the relation $\mu(C)$ varies with every material or phase.

‡ This statement is strictly accurate only for planar boundaries on which the number of particles remains constant. For example, adsorption phenomena or surface reactions require that one consider the concentration changes of the "surface phase."

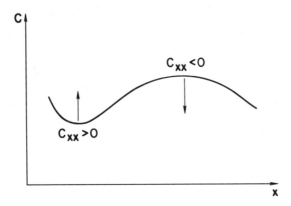

Fig 1.2: Profile of a "bumpy" concentration surface. It tends to smooth out with time.

C.1 Linearity of the diffusion equation

The diffusion equation satisfies a "principle of superposition": If C_1 and C_2 are two solutions of this equation and if α_1 and α_2 are two arbitrary constants, then $\alpha_1 C_1 + \alpha_2 C_2$ is also a solution, provided that

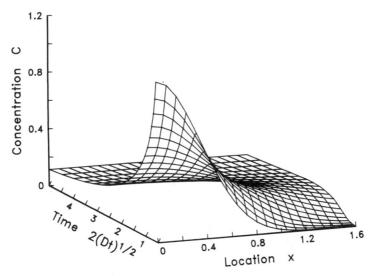

Fig. 1.3: Concentration distribution represented as a surface over the (x,t)-plane. This is, in fact, the gaussian (4.3) of Chapter 4. Note the essential singularity at $t = 0$.

the initial and boundary conditions of the full problem can be decomposed into those of the partial problems for C_1 and C_2. This is, in fact, an invariance property with respect to linear transformations of the dependent variable C. More generally, if $\mathscr{L}$ is a linear operator acting on a solution C, such as differentiation or integration with respect to the independent variables (x,t), or with respect to a parameter, then $\mathscr{L}\{C\}$ is also a solution. For example, the flux J, derived from a solution C, is also a solution of the diffusion equation. These facts are easy to verify directly on Eq. (1.14) because the operations in question are linear and they commute. Can you make similar statements for Eq. (1.15)?

C.2 Invariance under changes of scale and translations; symmetries

The diffusion equation does not change its form under reflection operations in space. More precisely, if $C(x,t)$ is a solution and if $x \to \bar{x} = -x$, then $C(\bar{x},t) = C(x,t)$ is also a solution, provided that the boundary conditions are unchanged under that transformation. This is easily verified (use the chain rule), because Eq. (1.14) is second order in x. Likewise, this equation is invariant under translations of the independent variables, *i.e.*, the addition of an arbitrary constant to x and t does not affect its form. A reversal of the arrow of time, however, does *not* preserve the form of the diffusion equation, because it is only first order in t. This is the mathematical content of "irreversibility," and diffusion is a prime example of an irreversible process. These results are amplified in the following

Exercise 1.5: *Affine transformations of the diffusion equation*

Consider transformations of the independent variables (x,t) of the form

$$\begin{pmatrix} \bar{x} \\ \bar{t} \end{pmatrix} = \begin{pmatrix} \alpha & \beta \\ \gamma & \delta \end{pmatrix} \begin{pmatrix} x \\ t \end{pmatrix} + \begin{pmatrix} x_0 \\ t_0 \end{pmatrix},$$

where all the coefficients α, β, γ, δ, x_0, and t_0 are constants. What conditions must these coefficients satisfy to preserve the form of the diffusion equation? ▮

There are many other interesting properties of the diffusion equation, among which the *maximum principle* and the proof of the *existence* and *uniqueness* of solutions. This last, in particular, is not merely of esthetic appeal. In practice, if we have found a solution to a particular diffusion problem — by whatever means — then we can be assured that we have found *the* solution to the problem in question. The interested reader will find a full account in Refs. [6,7], but the properties listed here are sufficient for our purposes. We shall become familiar with them in the following chapters.

1.4 NUMERICAL METHODS

In Sect. 1.2 we derived the diffusion equation (1.14) from a model (1.2) for diffusion. Now we do the reverse: Given the equation[†][§] $C_t = DC_{xx}$, we derive a so-called "finite-difference" form that can be solved numerically, *i.e.*, on a computer or even with a hand calculator. It is only required to reverse our steps. Whereas interpolation had provided the main tool, we now wish to *sample* the continuous distribution $C(x,t)$. Towards that end, introduce a (constant) rectangular grid on the domain of the independent variables (x,t) by choosing points

$$x_i = i\Delta x \quad \text{and} \quad t_n = n\Delta t \tag{1.21}$$

that are labeled by the integers i and n. This is shown on Fig. 1.4. The *a priori* arbitrary mesh sizes Δx and Δt determine the closeness of the finite-difference approximation to the exact solution, but, in general, Δx bears no relation to the jump distance a. Then, define a discrete function C_i^n that samples the continuous function $C(x,t)$ at the grid points (1.21), *i.e.*, $C_i^n = C(x_i, t_n)$, in a manner similar to Eq. (1.5). [Note, to avoid confusion, the superscript position for the time index n.] We now digress briefly on the approximation of derivatives.

It will be recalled that derivatives are defined in calculus through a limiting process on differences. For definiteness, consider the concentration profile $C(x)$ at a fixed instant of time, as shown schematically on Fig. 1.5. How does one then estimate the x-derivatives? As with

---

† The subscript notation for partial derivatives is handy; it saves both time and space. Thus here, for example, $C_t \equiv \partial C / \partial t$, and $C_{xx} \equiv \partial^2 C / \partial x^2$.

§ The continuum standpoint takes the view that the diffusion equation is a faithful representation of reality, although we know full well that a discrete random walk gives a more detailed description of diffusion.

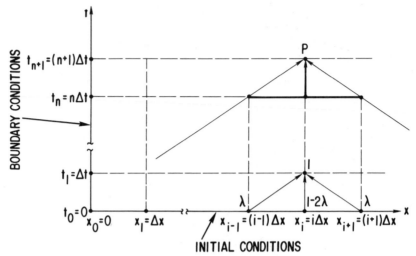

Fig. 1.4: Grid points necessary for finite-difference solutions. In the Euler method, evaluation at a current point P depends only on its three immediate predecessors.

Eqs. (1.6), a Taylor expansion around the current point x_i yields

$$C_{i\pm 1} = C_i \pm \Delta x \frac{\partial C}{\partial x} \Big|_{x_i} + \frac{1}{2}\Delta x^2 \frac{\partial^2 C}{\partial x^2} \Big|_{x_i} + O(\Delta x^3). \qquad (1.22a)$$

Hence, taking differences, we get the three possible representations (there are others; *e.g.*, see Exercise 1.7) for the first derivative:

$$\frac{\partial C}{\partial x} \Big|_{x_i} = \begin{cases} (C_{i+1} - C_i)/\Delta x + O(\Delta x), \\ (C_i - C_{i-1})/\Delta x + O(\Delta x), \\ (C_{i+1} - C_{i-1})/2\Delta x + O(\Delta x^2). \end{cases} \qquad (1.22b)$$

They measure the slope of the chords drawn in Fig. 1.5. The first alternative is called a *forward* (first) difference because the true derivative is estimated from forward data on the x-axis. The second is called a *backward* difference for similar reasons. Both of these are first-order accurate in Δx. The third alternative, the so-called *central* difference, yields a far better second-order accurate estimate, and it is clear from the figure that the corresponding chord is more nearly parallel to the tangent at the current point x_i.

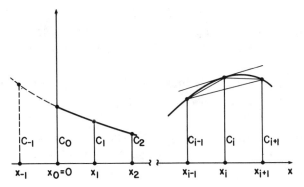

Fig. 1.5: Right-hand side: estimates of derivatives by chords. Left-hand side: a fictitious point x_{-1} for second-order accurate derivative estimation at a boundary.

Returning to the diffusion equation, we note with dismay that it is a *second* derivative in x that is required. But Eq. (1.22a) also provides a prescription for estimating C_{xx} that should be reminiscent of the right-hand side of Eq. (1.2), the so-called central second difference. Thus we have

$$\dot{C}_i(t) = (D/\Delta x^2)[C_{i+1}(t) + C_{i-1}(t) - 2C_i(t)] + O(\Delta x^2). \quad (1.22c)$$

But how do we estimate the time derivative? As with the space derivatives (1.22b), there are several answers to that question; each yields a distinct discretization scheme for the diffusion equation. Here we choose the simplest, the so-called "explicit (or Euler forward) scheme," in which the left-hand side of Eq. (1.22c) is estimated from data forward in time, *i.e.*, at t_{n+1}. A Taylor expansion in t, similar to Eq. (1.22a) and arrested at first order terms in Δt, then yields the desired finite-difference equation

$$C_i^{n+1} = C_i^n + \lambda(C_{i+1}^n + C_{i-1}^n - 2C_i^n), \quad (1.23a)$$

in which

$$\lambda = D\Delta t/\Delta x^2, \quad (1.24)$$

often called the *modulus*, is a dimensionless ratio of time to space mesh sizes. Equation (1.23a) represents the diffusion equation if we agree to neglect terms of order $O(\Delta x^2, \Delta t)$, the so-called local truncation error.

Exercise 1.6: *The second difference deserves its name*
For functions f_i over a discrete set of points, one defines [9,13] the first (forward) difference $\Delta f_i = f_{i+1} - f_i$. Show that the second difference operator $\Delta^2 f_{i-1}$, obtained by iterating the first difference (in complete analogy with second *derivatives*), has exactly the form in brackets of Eq. (1.22c). ∎

The finite-difference equation (1.23a) recursively generates a solution from initial data. Referring to Fig. 1.4, assume that the concentration is known at the initial time $t_0 = 0$ along the x-axis. This is known as the *initial condition*. Equation (1.23a), written for the time step $n = 0$, then *explicitly* gives the concentration distribution C_i^1 at the first time step $n = 1$ from three previous values of C. The computational pattern (or "computational molecule") has the shape of an inverted T with weights attached to its nodes, as shown on Fig. 1.4. The process can then be repeated for $n = 1, 2,...,$ and the solution evidently develops forward in time, from initial data, in lockstep with the above computational procedure. Such methods are therefore sometimes called "marching" procedures.

If the physical system that we are studying is finite in extent, then we must also be careful to respect the *boundary conditions* at each time step. For example, if the concentration is a known function $f(t)$ at the left boundary $x_L = 0$, say, corresponding to the index $i = 0$, then Eq. (1.23a) takes the form

$$C_1^{n+1} = (1-2\lambda)C_1^n + \lambda[C_2^n + f(t_n)] \tag{1.25a}$$

at the first interior point $i = 1$. Another important boundary condition arises when the gradient is known at the boundary, $\partial C/\partial x \big|_{x_L} = g(t)$. We could represent the gradient through the expression $(C_1^n - C_0^n)/\Delta x$, but the first Eq. (1.22b) shows that it is only first-order accurate in Δx, while the finite-difference equation (1.23a) is second-order accurate. A useful device to obtain better gradient conditions rests on the introduction of a fictitious point $i = -1$ outside the domain of the independent variables. This is shown on the left-hand side of Fig. 1.5.

The last of equations (1.22b) then demonstrates that $(C_1^n - C_{-1}^n)/2\Delta x$ is a second-order accurate estimate of the surface gradient. Next, we assume that the diffusion equation (1.14) and its finite-difference analog (1.23a) hold *at* the boundary, *i.e.*, we hope that its solution there is sufficiently smooth. In practice this expectation is not misplaced, except perhaps for the first time step (see Exercise 1.9 below). We thus have two equations from which to eliminate the fictitious quantity C_{-1}^n, and to get an expression

$$C_0^{n+1} = (1 - 2\lambda)C_0^n + 2\lambda[C_1^n - \Delta x \, g(t_n)] \tag{1.25b}$$

for the boundary values.

Exercise 1.7: *One-sided derivatives at boundaries*
The first derivative at a boundary can also be represented *without* the introduction of fictitious points. Consider, for example, the data at the three points $x_0 = 0$, x_1, and x_2 in Fig. 1.5. Interpolate the concentration by a second-degree polynomial in x and find its coefficients in terms of the data. Then estimate the derivative at the boundary $x = 0$. Is this a first- or a second-order estimate? ∎

The previous discretization scheme paid no attention to possible restrictions on the values of Δx and Δt, except for an intuitive sense that they should be "small." In other words, we expect the numerical solution of Eq. (1.23a) to converge towards the analytic solution of the diffusion equation if the mesh sizes are small. This can indeed be proved and is known as the *consistency* requirement of the finite-difference scheme [10,11]. In addition, we require that this convergence occur without spurious oscillations or overshoot. Thus we are also interested in the *stability* of the above finite-difference scheme. There are many ways to analyze this question: perhaps the most compelling is due to von Neumann. We will not consider this here because of space limitations and because the method is well documented elsewhere [9,11-13]. We observe, however, that Eq. (1.23a) cannot have oscillatory solutions if all its coefficients are of the same sign. Now the modulus λ, like D, is positive, and therefore the coefficient of C_i^n must also be positive. Hence, the modulus must obey the inequalities

$0 \leq \lambda \leq \frac{1}{2}$, and from the definition (1.24) we get the important restriction

$$\Delta t \leq \frac{1}{2} \frac{\Delta x^2}{D}. \tag{1.26}$$

And this is the crux of the matter, because we need a fine space mesh to resolve spatial gradients. Equation (1.26) then imposes an upper bound, *quadratic in* Δx, on the size of the time mesh. Small Δt's, in turn, may mean prohibitively large running times and possibly the accumulation of roundoff errors.

The explicit scheme (1.23a) is still very popular, in spite of its limitations. As mentioned earlier, there are several other ways to discretize Eq. (1.22c), even as there are several distinct finite-difference schemes that yield the same continuum equation (*cf.* Exercise 1.2). Some of these do not suffer from the limitation (1.26). An obvious candidate is the "fully implicit (or backward) scheme," in which the time derivatives of Eq. (1.22c) are estimated through $(C_i^n - C_i^{n-1})/\Delta t$, *i.e.*, the time analog of the second Eq. (1.22b). We then get the following set of linear equations

$$(1 + 2\lambda)C_i^n - \lambda(C_{i+1}^n + C_{i-1}^n) = C_i^{n-1} \tag{1.23b}$$

for the unknown quantities at time t_n. It can be shown that this scheme is unconditionally stable, *i.e.*, λ need not be restricted as in Eq. (1.26). The penalty for such freedom is the need to solve a (possibly large) linear system, and computer methods are definitely advisable. Fortunately, the coefficient matrix of Eqs. (1.23b) has a very simple structure, and is a prime example of a *sparse matrix*: Only the main, upper, and lower diagonals are filled; the rest is nothing but a bunch of zeros. There exist very powerful methods for the solution of such *tridiagonal* systems. These questions, as well as the discussion of other popular discretization schemes, such as that due to Crank and Nicholson, can be found in standard books on numerical analysis [9,11,13].†

---

† By now it should be obvious that the correspondence between continuum equations and their finite-difference analogs is not one-to-one.

Exercise 1.8: *Matrix form of the implicit method*
Cast Eqs. (1.23b) in matrix form, and observe that all the unknowns at time t_n must be obtained *simultaneously* from earlier data. Draw the computational molecule. Verify that the matrix is tridiagonal. How would you handle a boundary condition such as $C(0,t) = f(t)$? ∎

A few words regarding the information flow are in order. Referring back to Fig. 1.4 and to the computational pattern for the explicit scheme (1.23a), it is clear that the concentration's value C_i^{n+1} at the current point P depends on all previous values *inside* the sector (or pyramid, in more than one space dimension) with apex at P. Values outside this sector are irrelevant to the value at P. This "corporate" structure is, in fact, non-physical, because it implies that "signals" with velocity greater than $\Delta x/\Delta t$ (the inverse of the sector's inclination) cannot affect the current point's concentration.† On the other hand, the implicit scheme (1.23b) solves simultaneously for *all* concentration values along a time row n, and it uses all the information available below that row. Thus, it is both rather more democratic and more physical than the explicit scheme.

Numerical methods are very powerful because they are *not* limited to linear problems, typical of constant D. For example, the right-hand side of the nonlinear equation (1.15), if discretized, becomes $\Delta x^{-2}\{D_{i+1/2}[C_{i+1}(t)-C_i(t)] - D_{i-1/2}[C_i(t)-C_{i-1}(t)]\}$, a form that should remind us of Eqs. (1.1) and (1.3). The diffusivities must be evaluated at midpoints $i \pm 1/2$ to ensure an optimal truncation error. The reader is encouraged to write out such finite-difference schemes in both explicit and implicit form, and to think of ways to evaluate $D_{i \pm 1/2}$.

1.5 A NUMERICAL EXAMPLE

In electrostatics, a point charge at the origin creates the simplest potential distribution. Likewise for diffusion problems, one may ask how a unit mass influences an infinite medium if it was initially placed at the origin. This is called a *point source*, and it means, mathematically, that

---

† In contrast with the wave equation, it can be shown that the diffusion equation, being only first order in time, implies that signals must propagate with infinite speed.

the initial condition is $C(x,0) = \delta(x)$, or, in finite-difference terms, that[†]

$$C_i^0 = \delta_{i,0}, \quad i = ...,-2, -1, 0, 1, 2,... \tag{1.27}$$

This problem has no boundary conditions at any finite distance from the origin.[‡] If we now choose a value for the modulus λ, then we know from the last section that Eq. (1.23a) allows us to compute the concentration values at the first time step. We easily see from Fig. 1.6 and Eq. (1.23a) that only C_{-1}^1, C_0^1, and C_1^1 are nonzero. The process can be repeated, and we note that the point source yields nonzero concentration values within the sector of unit slope and of apex at the origin in the (i,n) plane.

The number of arithmetic operations for this simple problem (the so-called "operational count") can be roughly halved if we use its symmetry intelligently. This will be a first application of Property C.2 of Sect. 1.3, for both the diffusion equation and the boundary and

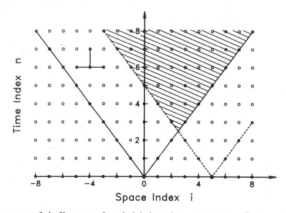

Fig. 1.6: Sectors of influence for initial point sources. Concentration values are additive in the hatched region. The computational molecule for Euler's method is shown.

[†] The Dirac delta-function $\delta(t)$ should be familiar from quantum mechanics. We will see more of it later on in Chap. 6. Its discrete analog is the *Kronecker delta* $\delta_{i,k}$, which has the value 1, if $i = k$, and 0 otherwise.

[‡] There *are* boundary conditions at infinity: the solution should decay sufficiently rapidly to ensure the convergence of sums over the concentration distribution, such as the total mass (see Appendix C).

initial conditions are symmetric under a reversal of the x-axis. This means that

$$C_i^n = C_{-i}^n, \tag{1.28a}$$

and we only need be concerned with the concentration values for positive index i, say. But what about the concentration along the line of symmetry $i = 0$? The answer is evident from Eq. (1.23a) on evaluating it at that position and taking Eq. (1.28a) into account:

$$C_0^{n+1} = (1-2\lambda)C_0^n + 2\lambda C_1^n. \tag{1.28b}$$

The complete solution is displayed in Table 1.1, which provides first a simple APL program, called $EXPLICIT$, and then some sample ouput. The arguments NF and L are the final value of the time index n (so that you won't run all night) and of the modulus λ, respectively. A few points are worth mentioning. First, we note that the output for $\lambda = 1$ is nonsensical because this case violates the stability criterion (1.26). The output for $\lambda = 0.5$ is somewhat better and is sometimes called the case of marginal stability. At least the concentration does not go negative. The cases $\lambda = 0.2$ and 0.1 are definitely more reasonable, and they also show the improvement that accrues with smaller moduli: Since $\Delta t = \lambda \Delta x^2/D$ by definition, it follows that t_n is proportional to λ for constant space mesh and diffusivity. Thus, for example, we compare directly the output rows $n = 1$ and 2 of the case $\lambda = 0.2$ with the rows $n = 2$ and 4 of the case $\lambda = 0.1$. Yet more refinement would indicate, heuristically at least, that the solution converges to something reminiscent of a gaussian. Note that stability is closely related to mass conservation, as is evident by summing concentration values over successive time-rows. The last case, $\lambda = -0.1$, is also of some interest, for it demonstrates the basic irreversibility of the diffusion equation. In fact, negative moduli would represent either a negative flow of time — and we saw that the diffusion equation is *not* invariant under time reversal — or that the diffusivity D is negative, which was excluded in Property B of Sect. 1.3, for then the diffusion process cannot be stable. This is a general feature of all irreversible processes: The choice of what one means by the *positive* flow of time is related to the positivity of the transport coefficient D.

Table 1.1: Listing and execution of an APL program
for a point source

```
       ∇ NF EXPLICIT L;N;C;C1
[1]    N←0
[2]    C←1,(1+NF)ρ0
[3]    'MODULUS IS ',⍕L
[4]    'I = ',(9 0)⍕0,ιNF
[5]  PRINT:'N = ',(⍕N),(9 3)⍕¯1↓C
[6]    C1←C[1]+2×L×C[2]-C[1]
[7]    C←C+L×(1⌽C)+(¯1⌽C)-2×C
[8]    C[1,NF+2]←C1,0
[9]    →(NF≥N←N+1)/PRINT
       ∇
```

```
       4 EXPLICIT 1
MODULUS IS 1
I =            0          1          2          3          4
N = 0       1.000       .000       .000       .000       .000
N = 1      ¯1.000      1.000       .000       .000       .000
N = 2       3.000     ¯2.000      1.000       .000       .000
N = 3      ¯7.000      6.000     ¯3.000      1.000       .000
N = 4      19.000    ¯16.000     10.000     ¯4.000      1.000
```

```
       4 EXPLICIT .5
MODULUS IS 0.5
I =            0          1          2          3          4
N = 0       1.000       .000       .000       .000       .000
N = 1        .000       .500       .000       .000       .000
N = 2        .500       .000       .250       .000       .000
N = 3        .000       .375       .000       .125       .000
N = 4        .375       .000       .250       .000       .063
```

```
       4 EXPLICIT .2
MODULUS IS 0.2
I =            0          1          2          3          4
N = 0       1.000       .000       .000       .000       .000
N = 1        .600       .200       .000       .000       .000
N = 2        .440       .240       .040       .000       .000
N = 3        .360       .240       .072       .008       .000
N = 4        .312       .230       .093       .019       .002
```

```
       4 EXPLICIT .1
MODULUS IS 0.1
I =            0          1          2          3          4
N = 0       1.000       .000       .000       .000       .000
N = 1        .800       .100       .000       .000       .000
N = 2        .660       .160       .010       .000       .000
N = 3        .560       .195       .024       .001       .000
N = 4        .487       .214       .039       .003       .000
```

```
        4 EXPLICIT -.1
MODULUS IS ⁻0.1
I =        0        1        2        3        4
N = 0   1.000     .000     .000     .000     .000
N = 1   1.200   ⁻.100     .000     .000     .000
N = 2   1.460   ⁻.240     .010     .000     .000
N = 3   1.800   ⁻.435     .036   ⁻.001     .000
N = 4   2.247   ⁻.706     .087   ⁻.005     .000
```

Exercise 1.9: *Two finite-difference calculations*

Write two programs using the explicit method to generate two solutions of the diffusion equation in a "half space" (a semi-infinite region) with the initial condition $C(x,0) = 0$ for all $x > 0$, and under the two boundary conditions: (a) $C(0,t) = C_0$ (a constant) and (b) $-DC_x(0,t) = J_0$ (also a constant). Both boundary conditions are valid for all $t > 0$. It is advised to sketch these conditions in the (x,t) plane, and to discuss what happens at the "infectious" origin. Use dimensionless variables, wherever possible. ∎

This is as good a point as any to introduce the notion of *Green's function*. We have just described the numerical solution to the finite-difference equation (1.23a) when the initial disturbance is a point source. Simply call that solution the Green's function (or "elementary solution") $G_{i,0}^n$, in which the second subscript 0 is to remind us that the point disturbance was placed at the origin.† Now, if that point source had been placed at any other point $x_k = k\Delta x$, then it is easily seen from the computational procedure that the solution would develop, exactly as before, inside the sector with apex at $i = k$, also shown (dashed) in Fig. 1.6. Thus $G_{i,0}^n = G_{i-k,k}^n$ has the translational symmetry‡ of this problem in accordance with Exercise 1.5: In an infinite region, the Green's function depends spatially only on the difference between the *observation point* i and the *source point* k. The second subscript k is thus superfluous. Further, the linearity of Eq. (1.23a) means that the effect of a unit source at $i = k$ simply adds to the effect of our initial source at the origin $i = 0$ (see again

---

† Strictly speaking, we have a family of functions G that depends on the modulus λ.

‡ Formally, G is invariant under the translation $i \rightarrow \bar{i} = i-k$, which implies $G_{i,0}^n = G_{\bar{i},0}^n = G_{i-k,k}^n$.

Property C.1 of Sect. 1.3). These effects are additive in the hatched region of Fig. 1.6, namely, in the intersection of the two sectors. We can now use these properties to compute the effect of an *arbitrary* initial condition $C(x,0) = \varphi(x)$, or its finite-difference analog

$$C_i^0 = \sum_k \varphi_k \delta_{i,k}. \tag{1.29a}$$

Again, because of the linearity property of the diffusion equation, each partial disturbance φ_k propagates independently, and each gives rise to a multiple of the Green's function for that source point k. The concentration at any time t_n is thus

$$C_i^n = \sum_k \varphi_k G_{i-k}^n, \tag{1.29b}$$

namely, a superposition of the effects of point sources that were distributed along the x-axis.

1.6 HIGHER DIMENSIONS AND COORDINATE SYSTEMS

The world is never simple and rarely one-dimensional. Sometimes, however, it can display a measure of symmetry. In this section we give, without much proof, the forms of Eqs. (1.12-15) for such cases. A thorough treatment would require knowledge of tensors [4,5,7].

Observe first that the concentration C is a scalar quantity. Thus, we expect the left-hand sides of Eqs. (1.13-15) to remain unchanged under coordinate transformations. The flux is another matter. In fact, the development in Sects. 1.1 and 1.2 indicates that it is a directed quantity, as is the partial derivative C_x. The following calculation is designed to demonstrate heuristically the vectorial nature of these quantities. Remember that our random-walk model was independent of a coordinate system, *i.e.*, the linear lattice can have an arbitrary orientation with respect to a fixed frame. This is shown schematically in Fig. 1.7. In that case, the lattice sites i are the end-points of position vectors $\mathbf{x}_i$. Taylor's expansion applied to the interpolating function $\hat{N}(\mathbf{x}_i,t)$ again leads to

$$N_{i\pm 1} = N_i \pm \mathbf{a} \cdot \nabla \hat{N} \big|_{\mathbf{x}_i} + O(a^2), \tag{1.30}$$

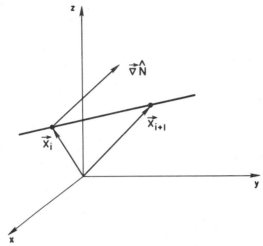

Fig. 1.7: Lattice points in space referred by their location vectors $\mathbf{x}_i$. The concentration gradient $\nabla \hat{N}$ need not be collinear with the lattice.

where now the jump *vector* $\mathbf{a} = \mathbf{x}_{i+1} - \mathbf{x}_i$ is dotted into the gradient $\nabla \hat{N}$.† Thus, with Eq. (1.1) for example, the flux $J_{i+1/2}$, which represents mass flow along the lattice in the direction of increasing i, is proportional to that scalar product. It is therefore related to the projection of a flux *vector* $\mathbf{J}$ onto the lattice. A more rigorous treatment (Appendix A) would show that Eq. (1.12) can be generalized to give

$$\mathbf{J}(\mathbf{x},t) = -D\nabla C \tag{1.31}$$

if it is further assumed that the diffusivity is isotropic.‡

Next, given that the flux is a vector quantity, we will show in Sect. 1.8 that the continuity equation (1.13) reads more generally

---

† Recall that the "del" or "nabla" operator ∇ has the form $\nabla = \mathbf{i}\,\partial/\partial x + \mathbf{j}\,\partial/\partial y + \mathbf{k}\,\partial/\partial z$ in cartesian coordinates (x,y,z) and that it serves to form the gradient ∇f and the laplacian $\nabla^2 f$ of a scalar function f. The divergence and curl of a vector field $\mathbf{v}$ are then $\nabla \cdot \mathbf{v}$ and $\nabla \times \mathbf{v}$, respectively.

‡ Isotropy means here that the concentration gradient is collinear with the flux. In crystals, for example, this may not be the case. Then Eq. (1.31) must be generalized to give $J^\alpha = -\Sigma_\beta D^{\alpha\beta}\partial C/\partial x^\beta$, where Greek superscripts label components (see Appendix A).

$$\frac{\partial C}{\partial t} = -\nabla \cdot \mathbf{J}. \qquad\qquad (1.32)$$

Finally, the last two equations, together, yield the diffusion equation

$$\frac{\partial C}{\partial t} = D\nabla^2 C \qquad\qquad (1.33)$$

if diffusion occurs in a region of more than one space dimension, and an obvious generalization of Eq. (1.15)

$$\frac{\partial C}{\partial t} = \nabla \cdot (D\nabla C) \qquad\qquad (1.34)$$

if the diffusivity is variable.

Now, a good many physical problems have, or are assumed to have, certain spatial symmetries. Just as it is never wise to fit a square peg into a round hole, we should take advantage of a body's symmetry to simplify the preceding Eqs. (1.31-34). Coordinate systems that mimic this symmetry are most helpful, mainly because the boundary conditions then assume simple forms. This will become evident in the following chapters. Of these, the cylindrical and spherical systems are the most commonly used. They are illustrated in Fig. 1.8, and we record below the expressions for the flux, divergence, and laplacian in these coordinates:

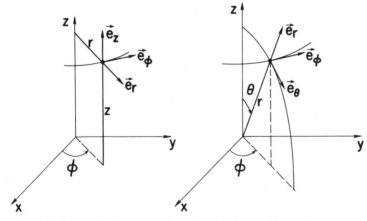

Fig. 1.8: Cylindrical (left) and spherical (right) coordinate frames and their unit vectors.

Cylindrical coordinates (r,φ,z) **Spherical coordinates** (r,ϑ,φ)

Coordinate transformations:

$x = r \cos \varphi$ $x = r \sin \vartheta \cos \varphi$

$y = r \sin \varphi$ $y = r \sin \vartheta \sin \varphi$ (1.35a)

$z = z$ $z = r \cos \vartheta$

Expressions for the flux $\mathbf{J} = -D\nabla C$:

$$J_r = -D\frac{\partial C}{\partial r} \qquad\qquad J_r = -D\frac{\partial C}{\partial r}$$

$$J_\varphi = -\frac{D}{r}\frac{\partial C}{\partial \varphi} \qquad\qquad J_\varphi = \frac{-D}{r \sin \vartheta}\frac{\partial C}{\partial \varphi} \qquad (1.35b)$$

$$J_z = -D\frac{\partial C}{\partial z} \qquad\qquad J_\vartheta = -\frac{D}{r}\frac{\partial C}{\partial \vartheta}$$

Expressions for the divergence $\nabla \cdot \mathbf{J}$:

$$\frac{1}{r}\frac{\partial}{\partial r}(rJ_r) + \frac{1}{r}\frac{\partial}{\partial \varphi}J_\varphi \qquad \frac{1}{r^2}\frac{\partial}{\partial r}(r^2 J_r) + \frac{1}{r \sin \vartheta}\frac{\partial}{\partial \varphi}J_\varphi$$

$$+ \frac{\partial}{\partial z}J_z \qquad\qquad + \frac{1}{r \sin \vartheta}\frac{\partial}{\partial \vartheta}(J_\vartheta \sin \vartheta) \qquad (1.35c)$$

Expressions for the laplacian $\nabla^2 C$:

$$\frac{1}{r}\frac{\partial}{\partial r}\left(r\frac{\partial C}{\partial r}\right) + \frac{1}{r^2}\frac{\partial^2 C}{\partial \varphi^2} \qquad \frac{1}{r^2}\frac{\partial}{\partial r}\left(r^2 \frac{\partial C}{\partial r}\right) + \frac{1}{r^2 \sin^2 \vartheta}\frac{\partial^2 C}{\partial \varphi^2}$$

$$+ \frac{\partial^2 C}{\partial z^2} \qquad\qquad + \frac{1}{r^2 \sin \vartheta}\frac{\partial}{\partial \vartheta}\left(\frac{\partial C}{\partial \vartheta}\sin \vartheta\right) \qquad (1.35d)$$

It must be emphasized that the components of the flux vector $\mathbf{J}$ are evaluated in the *orthonormal* frames† $(\mathbf{e}_r, \mathbf{e}_\varphi, \mathbf{e}_z)$ and $(\mathbf{e}_r, \mathbf{e}_\vartheta, \mathbf{e}_\varphi)$ of Fig. 1.8. Each unit $\mathbf{e}$-vector is tangent to a given coordinate line, and it points in the direction of increasing values of that coordinate.

---

† These components are called the *physical* components, to be distinguished from covariant and contravariant components that must be considered in non-orthonormal frames.

Problems in several space dimensions can be forbidding, as is illustrated by the full equations (1.35). Fortunately, a notable simplification accrues when problems are "rotationally symmetric," namely when the concentration depends only on the distance from a fixed axis or fixed point. Then, the angular and axial parts of the laplacian, Eqs. (1.35d), vanish identically, and the radial part appears to take a regular and simple form. This is indeed a general result:

Exercise 1.10: *Laplacian in rotationally symmetric systems*
If $C = C(r)$, where $r = (x^2 + y^2 + ...)^{1/2}$, and d is the number of dimensions of the rotationally symmetric subspace, show that

$$\nabla^2 C = \frac{1}{r^{d-1}} \frac{\partial}{\partial r}\left(r^{d-1} \frac{\partial C}{\partial r} \right) = \frac{\partial^2 C}{\partial r^2} + \frac{d-1}{r} \frac{\partial C}{\partial r}.$$

[Hint: Evaluate ∇^2 as the divergence of a gradient in rectangular coordinates.] ∎

It follows that Eqs. (1.32-34) take the form

$$\frac{\partial C}{\partial t} = -\frac{1}{r^{d-1}} \frac{\partial}{\partial r}\left(r^{d-1} J_r\right) = \frac{1}{r^{d-1}} \frac{\partial}{\partial r}\left(r^{d-1} D \frac{\partial C}{\partial r} \right) \qquad (1.36)$$

for such cases when the concentration has neither angular nor axial dependence. Finally, in the spherical case $d = 3$ it is easy to see that the function $u = rC$ satisfies the 1-d equation (1.14), with r substituted for x. This result is generalized in the following

Exercise 1.11: *Simplifications of Eq. (1.36)*
Assume that the concentration in a rotationally symmetric system has the form $C(r,t) = f(r)u(r,t)$. What can you say about the form of the functions f and u that simplify Eq. (1.36)? [Hint: Substitute and discuss the differential equations in f.] ∎

1.7 OTHER INSTANCES OF THE DIFFUSION EQUATION

A mere description of hopping on lattices would hardly justify a study of the diffusion equation. We hinted, however, that "conservation" or "balance" of certain physical quantities was the main issue. Although a complete treatment is outside our scope — it would require the ponderous arsenal of continuum mechanics — we now briefly consider other situations where the diffusion equation, in one guise or another, is a useful description of physical processes [14,15]. We write these in 1-d; the generalization to several dimensions is either trivial, as for heat conduction, or subtle, as for fluid dynamics.

A. Heat conduction in a stationary solid

This case is closest to mass diffusion. It differs, however, in that we must introduce *two* related transport coefficients. First, there is a continuity equation

$$\rho c \frac{\partial T}{\partial t} = -\frac{\partial J}{\partial x} \qquad (1.37a)$$

that expresses energy conservation. Here, T is the body's temperature, ρ and c are its mass density and heat capacity per unit mass, respectively, and J is the heat flux. The relation between temperature and flux is then often given by Fourier's law

$$J = -K \frac{\partial T}{\partial x}, \qquad (1.37b)$$

in strict analogy with Fick's first law (1.12) and with no greater degree of validity. It defines the *thermal conductivity K*. Together, these equations have the form (1.14) if we introduce the *thermal diffusivity* $\kappa = K/\rho c$, which indeed has the dimensions of D (cm^2/s). That Eq. (1.37a) represents energy conservation is now evident because $\rho c T$ is nothing but the internal energy density of a solid body. On the other hand, temperature is not a conserved quantity. Consequently, as mentioned above, two coefficients, K and κ, are required to describe temperature flow, rather than a single one, D, in the case of mass flow. The dimensions of J (W/cm^2) serve to find those of K (W/cm K).

B. Hydrodynamic flow past an infinite flat plate

Consider the following, sometimes known as "Stokes's first problem." A semi-infinite region, filled with some incompressible fluid at rest, is bounded by a horizontal plate normal to an x-axis that points into the fluid. If the plate suddenly begins to move in its plane, then its translational motion imparts a shear motion to the fluid, and momentum is

thus transferred. The fluid's velocity field has a single nonzero component v parallel to the plate, and one then shows that momentum conservation requires that it obey

$$\rho \frac{\partial v}{\partial t} = -\frac{\partial \tau}{\partial x}, \tag{1.38a}$$

where the shear stress τ serves to define the *viscosity* η if Newton's law

$$\tau = -\eta \frac{\partial v}{\partial x} \tag{1.38b}$$

holds. Note that the shear stress has the dimensions of a momentum
flux, *i.e.*, $[\tau] = (\text{g cm/s})/\text{cm}^2\,\text{s}$. Equations (1.38), together, again yield
an expression similar to Eq. (1.14) if one   defines the *kinematic
viscosity* $\nu = \eta/\rho$, which also has the dimensions of D. Again, the
momentum density ρv is the conserved quantity, rather than the velocity.

C. Quantum mechanics
Schrödinger's equation is also a diffusion equation, as is evident from

$$i\hbar \frac{\partial \psi}{\partial t} = \mathcal{H}\psi, \qquad \mathcal{H} = \frac{p^2}{2m} + V, \tag{1.39a}$$

and

$$p = -i\hbar \frac{\partial}{\partial x}. \tag{1.39b}$$

We now have an expression for the probability density ψ that is formally similar to a 1-d diffusion equation with a source† related to the
potential energy V, and where the momentum p plays the role of flux.
The analogy ends there, however, because the related diffusivity $i\hbar/2m$
is purely imaginary. Nonetheless, some methods of scattering theory are
formally similar to the solution by Green's functions of the diffusion
equation.

---

† The notion of distributed "sources" and "sinks" will be illustrated shortly.

Exercise 1.12: *Conservative form*
Rewrite Eqs. (1.37-39) in the form of Eqs. (1.12-13) when all coefficients are constants. Interpret the "concentration" and the "flux" in each case. What are the "diffusivities"? ∎

D. Semiconductor equations
The macroscopic transport of electrons and holes in semiconductors is also described by the same equations. If n and p denote their concentrations (*i.e.*, their *number* densities), then one shows that they are conserved in the sense of Eq. (1.13):

$$\frac{\partial n}{\partial t} = -\frac{\partial J_n}{\partial x}, \tag{1.40a}$$

$$\frac{\partial p}{\partial t} = -\frac{\partial J_p}{\partial x}. \tag{1.40b}$$

The fluxes, now, have a form more complex than (1.12) because both electrons and holes can be affected by electric fields E. This is a prime example of "drift" or "convective" terms in the flux. Because charged particles are constantly being accelerated between collisions, they achieve an average velocity, and hence give rise to a flow, that is proportional to the field. An elementary account will be given in Chap. 3. We shall find that the total fluxes are

$$J_n = -D_n \frac{\partial n}{\partial x} - \mu_n E n, \tag{1.40c}$$

$$J_p = -D_p \frac{\partial p}{\partial x} + \mu_p E p, \tag{1.40d}$$

where the *mobilities* μ_n and μ_p characterize the particles' response to the field. Note that the drift terms come with opposite signs to ensure that both mobilities are positive. This example also illustrates the possible complexities that arise in charged systems, for charges are also sources of the electric field. In fact, the three unknowns n, p, and E are related through Gauss's law of electrostatics

$$\varepsilon \frac{\partial E}{\partial x} = e(p - n), \tag{1.40e}$$

where e is the *magnitude* of the electronic charge and ε is the semiconductor's dielectric constant. The solution of any such problem is evidently highly nonlinear because of the products En and Ep in Eqs. (1.40c,d). Ionic transport in solutions or in dielectrics is also governed by similar equations.

E. Current flow through a conductor

Free electrons in a conductor are neither created nor destroyed. It can be shown that their density n obeys Eq. (1.40a) above, but that the diffusive component of their flux J_n is negligible compared to the drift component. If $\rho = -en$ represents the electron *charge density* and if it is recalled that the electric field derives from an electrostatic potential ϕ (such that $E = -\phi_x$), then we have a continuity equation

$$\frac{\partial \rho}{\partial t} = -\frac{\partial J}{\partial x} \tag{1.41a}$$

that requires a current density

$$J = -eJ_n = -\mu_n en \frac{\partial \phi}{\partial x}. \tag{1.41b}$$

This last is nothing but Ohm's law, with *conductivity* $\sigma = \mu_n en = -\mu_n \rho$. The potential, however, satisfies Laplace's equation $\phi_{xx} = 0$ (in 1-*d*) because the free electrons in a conductor are compensated by a background sea of positive charges. The right-hand side of Gauss's law (1.40e) thus vanishes. These problems are linear, in contrast to semiconductor problems, because finding the field distribution is decoupled from the calculation of current density.

1.8 GENERAL BALANCE LAWS AND CONSTITUTIVE RELATIONS

There are many other manifestations of the diffusion equation: in biology, in queueing theory, and in the description of telephone traffic and other bottlenecks. At this point, the conclusion of the first chapter, we might ask why this is so. The answer rests on the observation that there are quantities in nature that are conserved or, at least, that can be balanced in some grand accounting scheme [15,16]. Such a quantity — call it Q — is additive in the sense that its value for a collection of subsystems is the algebraic sum of its values for each subsystem. Examples include the mass, charge, momentum, angular momentum, various forms of energy, and entropy of any macroscopic system. One then assumes that matter is continuous and that there exists a density function, or *field q* (per unit mass)† such that

$$Q(t) = \int_{\mathscr{D}} \rho(\mathbf{x},t)\, q(\mathbf{x},t)\, dV. \tag{1.42}$$

Thus, an additive quantity is expressed as an integral over a domain $\mathscr{D}$ (a specified open and connected region of space) that contains a given physical system. Here, ρ is again the mass density, itself a field. In former days, additive quantities were known as "extensive," probably to distinguish them from fields, which were known as "intensive" quantities. Some books make much of this distinction, but in fact, Eq. (1.42) is all there is to say.

To begin our balancing act requires merely that we distinguish what causes Q to change with time. There are exactly two reasons for change: either because there are internal sources or sinks σ distributed inside the system,‡ or because there are exchanges, represented by a vector field $\mathbf{J}_{\text{tot}}$, with the world at large. The first are internal to $\mathscr{D}$, while the second act through its closed boundary $\partial\mathscr{D}$, as shown in Fig. 1.9. Thus we have the ledger

$$\dot{Q} = \int_{\mathscr{D}} \sigma\, dV - \int_{\partial\mathscr{D}} \mathbf{J}_{\text{tot}} \cdot \mathbf{n}\, dA,$$

where $\mathbf{n}$ is the outward pointing normal to the system's boundary and dA is its area element. It is now a simple matter to use the definition (1.42), to carry out the differentiation, and to use the divergence theorem

---

† A field is generally understood to be *any* (tensor-valued) function of the space and time variables $(\mathbf{x},t)$. Not all fields, however, are density functions, *i.e.*, functions such that their integral over all space has a physical meaning. For example, the temperature is a field, but its integral over space means nothing at all.

‡ The notion of distributed source (or sink) is best illustrated through an example: If a binary gas undergoes a homogeneous chemical reaction $A \rightarrow B$, then the mole density of each species changes as a function of time. The reaction *rate* is then a source-like term, because it describes the removal of moles of A from the left side of the above reaction and the creation of moles of B on its right-hand side.

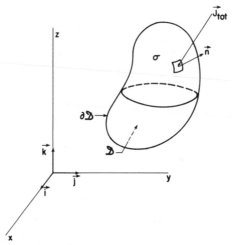

Fig. 1.9: The domain $\mathcal{D}$ of any physical system, its boundary $\partial\mathcal{D}$, and the unit exterior normal **n**. The flux and (internal) source fields are shown.

$$\int_{\mathcal{D}} \nabla \cdot \mathbf{J}_{tot}\, dV = \int_{\partial\mathcal{D}} \mathbf{J}_{tot} \cdot \mathbf{n}\, dA. \qquad (1.43)\dagger$$

We then find‡

$$\frac{\partial \rho q}{\partial t} = -\nabla \cdot \mathbf{J}_{tot} + \sigma, \qquad (1.44)$$

an equation already reminiscent of the continuity equation (1.32). It remains only to recognize that a given field q can also be carried (or "convected") by an overall fluid motion **v**, so that the decomposition of the total flux

$$\mathbf{J}_{tot} = \rho q \mathbf{v} + \mathbf{J} \qquad (1.45)$$

---

† This theorem is sometimes attributed to Green or to Ostrogradskii; it was surely known to Gauss. It also illustrates a minor semantic difficulty associated with the word "flux:" In potential theory, one often defines the flux of a vector field as an expression similar to the right-hand side of Eq. (1.43); it corresponds to our *integrated* flux. Some authors thus prefer the word "current" for the vector quantity **J** representing the flow of a physical quantity q.

‡ The derivation rests critically on the assumption that $\mathcal{D}$ be continuous. If not, as in a system of several phases, the domain must be broken up into disjoint pieces and each integral considered separately, much as was done for Properties A.2-3 of Sect. 1.3.

defines the true flux **J** due to causes other than convection. In particular, the total mass is an additive quantity with density ρ. Moreover, mass can be neither created nor destroyed, and its total flux is merely that due to convection. Thus, putting $q = 1$, $\sigma = 0$, and $\mathbf{J}_{tot} = \rho\mathbf{v}$ in the above Eq. (1.44), we get the (mass) continuity equation

$$\frac{\partial \rho}{\partial t} = -\nabla \cdot (\rho\mathbf{v}). \tag{1.46}$$

Inserting expression (1.45) into Eq. (1.44) and using Eq. (1.46), we obtain the general balance equation for any field†

$$\rho\frac{\partial q}{\partial t} + \rho\mathbf{v} \cdot \nabla q = -\nabla \cdot \mathbf{J} + \sigma. \tag{1.47}$$

This is the point of departure for a full development and justification of the special cases, and others as well, that were mentioned in the previous section. Note that Eq. (1.47) is first order in time and that it contains the continuity equation (1.32) as a special case when the velocity field and source term vanish.

Here, we shall be content to note that the balance equation (1.47) is an exact statement in the sense that all one does is to distinguish causes and effects. Nevertheless, for a given source σ and impressed velocity field **v**, that equation relates the *two* unknowns: field q and flux **J**. To solve such an equation still requires a relation between these unknowns. Such a relation is called "constitutive" or "phenomenological" because it depends on the system's basic building blocks and because it must ultimately find its justification in a microscopic model of the transport process. For example, our random walk for mass diffusion was precisely such a finer-scale model. Indeed, it yielded a *linear* relation between concentration gradient and mass flux, and that relation, Fick's first law (1.12) or (1.31), was characterized by a single *transport coefficient*, the diffusivity, which was also estimated in terms of jump frequencies and jump distances. Experimental observation shows that such linear relations are not limited to mass transport. The laws of Fourier (1.37b), Newton (1.38b), and Ohm (1.41b) are often obeyed when systems are "close to equilibrium." The following heuristic argument may serve.

---

† This calculation requires an important result of vector analysis: $\nabla \cdot (\rho q\mathbf{v}) = q\nabla \cdot (\rho\mathbf{v}) + (\rho\mathbf{v}) \cdot \nabla q$.

At equilibrium, the various fields q that characterize the state of a given physical system are generally (but not always) independent of the space coordinates (x,y,z). The fluxes **J** of these fields also vanish at equilibrium. Therefore, when a system is disturbed, it is plausible that the flux **J** of any such field will be functionally related to the *gradients* of q. A Taylor expansion around the equilibrium state then yields†

$$\mathbf{J} = -D\nabla q, \tag{1.48}$$

which defines the transport coefficient D. This relation has an immediate interpretation, for the equation $q(\mathbf{x}) = $ const defines level surfaces (isoconcentration surfaces, isotherms, equipotentials, *etc.*). The corresponding fluxes are then orthogonal to these surfaces if the medium is isotropic (D is then a scalar). The gradients ∇q are sometimes called "thermodynamic forces" because they appear to cause the corresponding fluxes or currents **J**, and, in the jargon of electrical circuits, the transport coefficient D has the character of a conductance.‡ Finally, the minus sign and the positivity of D in Eq. (1.48) are related to the stability of transport processes, much as in our elementary discussion in Sects. 1.3 and 1.5. Contemporary folklore couches these considerations in terms of "minimum entropy production" or "dissipative structures." For our purposes, however, these would be mere slogans, and we shall merely say that physical systems evolve, most often, towards states of lower energy. In a nutshell, *physical quantities coast downhill all the way* [17].

Exercise 1.13: *Invariance of balance laws under affine transformations*

Consider, for simplicity, the 1-*d* form of Eqs. (1.47-48). Eliminate the flux in favor of the field to get a "diffusion equation," and recall the

---

† The total flux (1.45) cannot be thus functionally related because the resulting phenomenological relation would not be galilean invariant, a requirement of many physical laws.

‡ By now it should be clear that linear constitutive relations such as Eq. (1.48) do not have the same degree of generality as the balance laws (1.47). There exist physical systems and phenomena where the former do not hold, *e.g.*, viscoelastic and porous media, hysteresis, and spinodal decomposition. Furthermore, multicomponent systems can exhibit "uphill" diffusion, so that the diffusivity appears to be negative in certain regions. Although the system's energy can increase locally in these regions, the overall energy must decrease with time.

affine transformation of Exercise 1.5. What conditions must the transformation coefficients satisfy to preserve the form of these equations? It is assumed that the field q is a scalar under that transformation. How must the source σ, the diffusivity D, and the velocity v transform? ∎

REFERENCES

1. Lord Rayleigh, "On James Bernoulli's Theorem in Probabilities," *Phil. Mag.* **47**, 246 (1899). [Reprinted in *Scientific Papers* **4**, p. 370 (Dover, New York 1964).]

2. S. Chandrasekhar, "Stochastic Problems in Physics and Astronomy," *Revs. Mod. Phys.* **15**, 1 (1943). [Reprinted in *Selected Papers on Noise and Stochastic Processes*, N. Wax, Ed. (Dover, New York 1954).]

3. P.G. Shewmon, *Diffusion in Solids* (McGraw-Hill, New York 1963).

4. C.C. Lin and L.A. Segel, *Mathematics Applied to Deterministic Problems in the Natural Sciences* (MacMillan, New York 1974).

5. H.S. Carslaw and J.C. Jaeger, *Conduction of Heat in Solids*, 2nd ed. (Oxford University Press, London 1959).

6. D.V. Widder, *The Heat Equation* (Academic Press, New York 1975).

7. A.N. Tikhonov and A.A. Samarskii, *Equations of Mathematical Physics* (Pergamon Press, Oxford 1963).

8. R. Ghez and W.E. Langlois, "More on the Concentration Dependence of Fick's Laws," *Amer. J. Phys.* **54**, 646 (1986).

9. B. Carnahan, H.A. Luther, and J.O. Wilkes, *Applied Numerical Methods* (John Wiley, New York 1969).

10. R. Courant, K. Friedrichs, and H. Lewy, "Über die Partiellen Differenzengleichungen der Mathematischen Physik," *Math. Ann.* **100**, 32 (1928). [Translated and reprinted in *IBM J. Res. & Dev.* **11**, 215 (1967).]

11. R.D. Richtmeyer and K.W. Morton, *Difference Methods for Initial Value Problems*, 2nd ed. (Interscience, New York 1967).

12. G.G. O'Brian, M.A. Hyman, and S. Kaplan, "A Study of the Numerical Solution of Partial Differential Equations," *J. Math. & Phys.* **29**, 223 (1950).

13. E. Isaacson and H.B. Keller, *Analysis of Numerical Methods* (John Wiley, New York 1966).

14. R.B. Bird, W.E. Stewart, and E.N. Lightfoot, *Transport Phenomena* (John Wiley, New York 1960).

15. J.C. Slattery, *Momentum, Energy, and Mass Transfer in Continua*, 2nd ed. (Robert E. Krieger Publ. Co., New York 1981).

Chapter 2

STEADY-STATE EXAMPLES

This chapter introduces the notion of steady state through several physically significant examples.   Indeed, there exist many time-dependent problems, among which are problems of phase growth, whose solution can be "almost" at steady state.   Such is the case for oxidation kinetics, precipitate growth, and crystal growth under conditions of constant supersaturation. Along the way, the equilibrium properties of binary systems and of "small" phases will be reviewed.

2.1 THE STEADY STATE IS *NOT* THE EQUILIBRIUM STATE

The steady state must be carefully distinguished from the equilibrium state, even though the latter is, in a sense, a subcase of the former. Recall first that equilibrium obtains when the total entropy of a body is maximum under certain constraints (*e.g.*, constant energy, volume, or masses of various species) that characterize macroscopically a given physical system.   With this variational principle, one then *proves* that the fields "conjugate" to the constraints, such as temperature, pressure, chemical potentials, *etc.*, are independent of the space variables within each phase.   These fields are said to be *constant*. On the other hand, we have the

Definition: Any field q is in a steady state if $\partial q / \partial t = 0$.

Therefore, a steady-state field q can depend only on position $\mathbf{x}$ and on any other coordinates that characterize the physical system's internal state: Measurements, today, of a steady-state quantity yield the same value as any later measurement, as long as all external conditions remain constant in time.

Our simple random walk will serve to illustrate this difference. If $\dot{N}_i = 0$, $\forall i$,† then Eq. (1.3) shows that the fluxes $J_{i\pm1/2}$ are equal regardless of the site i; they equal a constant J_0, say. In addition, Eqs. (1.1) are now simple difference equations whose solution is evidently $N_i = \alpha - (2J_0/\Gamma)i$, where α is another constant. The flux is thus constant with respect to the space coordinate i, and the concentration is a linear function of that coordinate. This is characteristic of the steady state in one dimension (for constant Γ's). In contrast, equilibrium is that subcase where the flux is not only constant but strictly vanishes, which implies that the particle distribution is then constant.‡

Exercise 2.1: *Steady-state thermal conduction through a composite slab*
Assume constant conductivity K_i in each distinct region (x_{i-1}, x_i), $i = 1,2,...,M$, and temperature continuity at the phase boundaries. The temperatures at the external boundaries x_0 and x_M are given. Find the temperature distribution and interpret the result in terms of "thermal resistances," "current," and "potential drop." ∎

It follows from the definition and from the general balance law (1.47) that the steady-state values of any density function q must satisfy

$$\rho \mathbf{v} \cdot \nabla q = -\nabla \cdot \mathbf{J} + \sigma. \tag{2.1}$$

If, in addition, the velocity field and the source term are negligible, then it follows that the flux $\mathbf{J}$ must be divergence free:

$$\nabla \cdot \mathbf{J} = 0. \tag{2.2a}$$

If, further, field and flux are related by the constitutive relation (1.48)

$$\mathbf{J} = -D\nabla q, \tag{2.2b}$$

repeated here for convenience, then the field q must satisfy

---

† Recall the handy notation from logic: $\forall a$ and $a \in b$, which read "for all a" and "a is a member of b," respectively.

‡ In general, equilibrium can exist even if the concentration distribution is nonconstant, because only the constancy of the chemical potential is required. This will be illustrated in the next chapter with a short introduction to double layers.

$$\nabla \cdot (D\nabla q) = 0. \tag{2.2c}$$

Finally, if the diffusivity is independent of location, then the field q obeys Laplace's equation

$$\nabla^2 q = 0. \tag{2.2d}$$

These are the equations that we intend to illustrate and solve in this chapter.

Exercise 2.2: *Steady state solutions when D = D(C)*
Find all physical solutions $C(x)$ for steady-state diffusion in a half space, without sources or convective motion, if $C(0) = C_0$ and $D = D_0(C/C_0)^\nu$, where C_0 and D_0 are arbitrary positive constants and the exponent ν is arbitrary. ∎

2.2 GENERAL SOLUTIONS FOR ROTATIONALLY SYMMETRIC CASES

When the concentration depends only on the distance to a fixed point or axis, Eq. (1.36) shows that Eqs. (2.2a,b) reduce to

$$\frac{d}{dr}\left(r^{d-1}J\right) = 0, \tag{2.3a}$$

$$J = -D\frac{dC}{dr}, \tag{2.3b}$$

where, for brevity, the subscript r of the radial flux component has been dropped. [Recall that d is the dimension of the rotationally symmetric subspace.] The first equality (2.3a) yields immediately

$$J = \beta/r^{d-1} \tag{2.4a}$$

for all points other than the origin. With this "first integral," the second equality (2.3b) gives

$$C = \begin{cases} \alpha - [\beta/D(2-d)]\, r^{2-d} & \text{for } d \neq 2, \\ \alpha - (\beta/D)\ln r & \text{for } d = 2, \end{cases} \tag{2.4b}$$

if, for whatever reason, the diffusivity is spatially constant, *i.e.*, $\partial D/\partial r = 0$. Here, α and β are integration constants. Note also that Eqs. (2.4) are singular at the origin $r = 0$, except in the case of one space dimension.

We now apply these formulas to a particularly simple, yet useful case: *diffusion through a long hollow cylinder.* This method is of practical interest since it has often been used in the past to measure diffusivities or thermal conductivities [1,2], even when these depend on concentration or temperature. Consider, therefore, a long circular cylinder of length H and of radius R_2 made of some material whose diffusive properties we wish to investigate. Then bore a hole of radius R_1 along its axis, as shown schematically in Fig. 2.1. This has two functions: To remove physically the singularity of the previous equations, and to allow the passage of a substance whose diffusive properties through the cylinder we wish to investigate. For example, for a metal cylinder, one could pass a reactive substance such as water to study the cylinder's oxidation kinetics, or an impurity to modify its electrical or mechanical properties. Assume for simplicity that the concentration of the diffusing species is fixed at the inner and outer cylindrical surfaces, namely, $C(R_1) = C_1$ and $C(R_2) = C_2$. We also assume that the cylinder is so long that the concentration does not depend appreciably on axial position z. Consequently, the concentration distribution $C(r)$ is purely

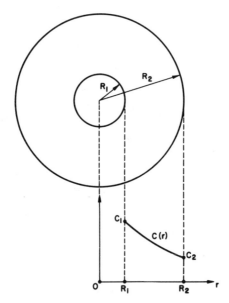

Fig. 2.1: Schematic representation of the hollow-cylinder problem.

radial, and the previous relation (2.4a) holds in each transverse section of two dimensions. A short discussion of this problem follows.

Applying Eq. (2.4a) for $d = 2$, we get $rJ(r) = \text{const } \forall r$. Therefore, the total amount of substance that passes, in a time interval t, through *any* cylindrical surface of radius r and height H is

$$M = 2\pi rHtJ. \tag{2.5a}$$

Note that M has the sign of J and that it has dimensions of grams if $[J] = \text{g/cm}^2$ s. Now for a quick aside. This result appears to contradict the divergence theorem (1.43) applied to the flux vector **J**: By virtue of Eq. (2.2a) [or the equivalent (2.3a)], that vector ought to be divergence free, and yet its integral over cylindrical surfaces is here proportional to the right-hand side of Eq. (2.5a). Thus, whereas blithe application of this theorem would predict no integrated flux, we find a quantity proportional to M which is generally nonzero. The resolution of this "mini-paradox" rests on the simple observation that the divergence theorem is true for continuous regions $\mathscr{D}$ only, *i.e.*, for regions whose boundaries $\partial \mathscr{D}$ *do not enclose singularities*. Here, in the derivation of Eq. (2.5a), we have enclosed a line of singularities, namely the cylinder's axis $r = 0$.

Equation (2.5a) leads immediately to two useful results if we remember Eq. (2.3b). Then,

$$M = -2\pi rHtD\frac{dC}{dr} \tag{2.5b}$$

can be viewed either as an equation for D

$$D(r) = -\frac{M}{2\pi rHt\, C'(r)}, \tag{2.6a}$$

or as a differential equation for C. In the first case, measurements (*e.g.*, by sectioning) of a concentration profile $C(r)$, under conditions of known constant mass flow rate $\dot{M}$, give the position dependence of the diffusivity in the annular region $R_1 < r < R_2$. From the measured profile we can then construct the dependence $D(C)$. In the second case, variables separate in Eq. (2.5b) to give

$$M \ln(R_2/R_1) = 2\pi Ht \int_{C_2}^{C_1} D(C)\, dC. \tag{2.6b}$$

Since the above integral is nothing more than the average diffusivity $\overline{D}$ times the concentration drop $(C_1 - C_2)$ over the annulus, Eq. (2.6b) provides a rapid estimate of the diffusivity if we can measure all other quantities accurately. The computation of the concentration distribution is the subject of the next

Exercise 2.3: *Concentration distribution in a long hollow cylinder*
Assume that $\partial D/\partial C = 0$. Find the function $C(r)$ either by determining the constants in Eq. (2.4b) or by integrating the equivalent Eq. (2.5b). Discuss the case of a "thin" annulus. [Hint: Use dimensionless variables such that $(r - R_1)/R_1 \ll 1$.] ∎

2.3 THE THERMAL OXIDATION OF SILICON

In the semiconductor industry, there is perhaps no single processing step more important than the formation of SiO_2 on silicon. As is well known [3], these overlayers form protective "masks" against the diffusion of impurities, and they also provide electrical insulation and dielectric properties necessary for the fabrication and operation of semiconductor devices. These facts, alone, warrant the study of SiO_2 growth kinetics. From our point of view, however, this physical system is ideally suited to show how one can develop a physical model: what assumptions must be retained and which ones to discard, how to set up the problem mathematically, and how to test the validity of the model.

First, some experimental evidence. Silicon is a highly reactive element. In particular, a stable amorphous oxide forms on its surface if it is exposed to an oxygen-bearing ambient. This can be, typically, a mixture of inert gases (*e.g.*, H_2 or N_2) with O_2 or H_2O. A thin (~ 10 Å) layer of oxide then forms, even at room temperature.† The oxidation reaction is, however, an *activated process*, which means that it can be greatly accelerated at elevated temperatures ($\gtrsim 700°C$) to promote the formation of much thicker oxide layers, which are then useful. These layers are also *conformal* and *adherent*, in the sense that

---

† This so-called "native" oxide causes processing engineers much grief because it covers what should be an otherwise clean Si surface. The considerations of this section do *not* apply to the formation of native oxides.

they follow exactly the topography of the underlying Si substrate. Other shreds of experimental evidence include the following: The oxidation *rate* is not a constant; it decreases roughly as $t^{-1/2}$ for long enough oxidation times, and it is approximately proportional to the oxygen partial pressure p. We anticipate that the oxidation process is mainly due to the migration of *oxygen molecules* to the Si/SiO$_2$ interface where they react, at the silicon substrate's expense, to form fresh SiO$_2$. Silicon does *not* appear to migrate significantly. These observations form the basis of an oxidation model due to Deal and Grove [4], which we now develop in 1-d.

Figure 2.2 shows schematically a planar layer of SiO$_2$ during its formation. It is sandwiched between an O$_2$ ambient, with partial pressure p, and the Si substrate. This is clearly a three-phase system and a first example of a *phase change*. We seek an expression for the oxide's thickness L as a function of time and of other experimentally controllable parameters. We assume that:

a. Oxygen dissolves molecularly at a rate J_0 into the existing SiO$_2$ at the outer interface (taken as the origin of coordinates), and its concentration there is given by Henry's law

$$C_0 = Kp, \tag{2.7a}$$

where the *solubility* K depends only on temperature T.

b. Oxygen diffuses molecularly through the existing SiO$_2$, and it does not react *homogeneously*† with the glass network. This assumption has been verified experimentally [5]. If $C(x,t)$ is the O$_2$ concentration *in excess* of what is contained in the network, then we assume that its evolution can be described by the diffusion equation (1.14).

c. Oxygen molecules react *heterogeneously* at the inner interface according to the overall reaction

---

† Recall that homogeneous reactions occur *inside* a given phase. They are described mathematically by source terms in balance laws, and they characterize chemical transformations at interior points. On the other hand, heterogeneous reactions occur at phase *boundaries*. They are described mathematically by boundary conditions. Here, the reaction (2.7b) occurs at the Si/SiO$_2$ interface, assumed to be a geometric surface and not a 3-d transition region. Recent measurememments [6] indicate that this may be but a useful idealization. Note also that the "back-reaction" and equilibrium concentration in Eqs. (2.7b,c) are assumed negligible.

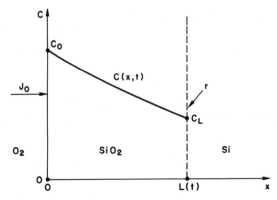

Fig 2.2: Three-phase oxidation system and excess O_2 distribution in the oxide.

$$Si + O_2 \rightarrow SiO_2. \tag{2.7b}$$

The chemical *reaction rate r* must thus be first order in the oxygen concentration C_L that exists at the Si/SiO_2 interface. Expressed mathematically,

$$r = kC_L, \tag{2.7c}$$

i.e., the reaction rate is linear in the O_2 concentration and the rate constant k depends only on temperature. Note that r has the dimensions of flux and that the position of the Si/SiO_2 interface $L(t)$, in our coordinate system, is also the instantaneous oxide thickness.

d. Diffusion occurs at steady state, so that the diffusion equation reduces simply to $C_{xx} = 0$. This statement clearly requires elaboration because oxidation is evidently a time-dependent process (the oxide thickness L is a nonlinear function of time). In fact, this example provides us with a first nontrivial problem: How can we model a physical system whose boundary moves? An complete answer, in this instance, requires the numerical solution of the full time-dependent problem. It is useful, however, to assume that the concentration field C is approximately at steady state, namely that

$$C \cong \alpha + \beta x, \tag{2.7d}$$

in accordance with the results of Sect. 2.1, but where the "constants" α and β can depend weakly on time. This is sometimes called the assumption of *quasi*-steady state, to which we will return later on.

To solve this problem demands that we take mass conservation somewhat more seriously, because the inner interface $x = L(t)$ is the seat of physical processes other than the mere exchange of mass by diffusion. We must thus refine the reasoning implicit in Properties A of Sect. 1.3. For definiteness, let us say that all concentrations are measured in units of number densities, i.e., $[C] = cm^{-3}$. Then the total number† of excess O_2 molecules in the system, per unit of substrate area, is

$$M_{O_2}(t) = \int_0^{L(t)} C(x,t)\, dx. \tag{2.8a}$$

Its time derivative must be equal to any known reason for change, namely, an accretion by the inward flux J_0 minus a loss due to the reaction rate r of Eq. (2.7b). Since we have no reaction at the outer interface that changes the number of moles (such as in the case of dissociative adsorption; see Exercise 2.6 below), it follows that the flux must be continuous there, exactly as in Eqs. (1.19). Carrying out the indicated differentiation as in Eq. (1.20) and balancing the fluxes at $x = 0$, we find the condition

$$D\frac{\partial C}{\partial x}\Big|_L + r + C_L\dot{L} = 0 \tag{2.8b}$$

that must hold at the inner interface. It remains to relate the reaction rate to the movement of that interface. This is easily done by considering the total number of SiO_2 molecules in our system. The concentration C_{SiO_2} is evidently a constant and is the reciprocal of the molecular volume Ω of SiO_2. Thus

$$M_{SiO_2}(t) = \Omega^{-1}L(t), \tag{2.8c}$$

whose time derivative can only be the gain of SiO_2 though the right-hand side of the reaction (2.7b). Thus, we get the desired relation

$$r = \Omega^{-1}\dot{L}. \tag{2.8d}$$

---

† It is assumed here that no oxygen dissolves in the Si substrate. Recent studies have shown that this assumption is open to question, as is the matter of Si dissolution into the oxide and into the substrate itself (as self-interstitials).

We are now almost done with model building. It simply remains to solve the above equations. With Eq. (2.7c), eliminate r from Eqs. (2.8b,d):

$$DC_x \big|_L + kC_L(1 + \Omega C_L) = 0, \tag{2.9a}$$

$$\dot{L} = \Omega k C_L. \tag{2.9b}$$

These are boundary conditions that must hold at the Si/SiO$_2$ interface. The first determines the as yet unknown concentration C_L at that interface, and the second, a Stefan condition, gives that interface's rate of motion. On the other hand, the assumed steady-state distribution (2.7d) can also be written

$$C(x) = C_0 - (C_0 - C_L)x/L \tag{2.10}$$

in terms of the concentrations at both interfaces. Inserting this expression into Eq. (2.9a) gives a quadratic equation for C_L, and the problem would be vastly simpler if we could neglect the second term in the parenthesis of Eq. (2.9a). This is indeed the case, as the following numerical estimate shows.

The atomic volume of SiO$_2$ is equal to its molecular mass (60.1 g/mole) divided by its density (2.27 g/cm^3) and by Avogadro's number (6.02 × 10^{-23} mole^{-1}). Thus, $\Omega = 4.4 \times 10^{-23}$ cm^3. On the other hand, the concentration C_L is certainly less than C_0 that is imposed at the gas/SiO$_2$ interface. Now, one finds experimentally that the solubility K in Eq. (2.7a) is of the order 7.5×10^{16} atm^{-1} cm^{-3}, so that the term we are considering, ΩC_L, is less than $\Omega C_0 = 3.4 \times 10^{-6}$ at a pressure of one atmosphere. This is definitely negligible in comparison with unity in Eq. (2.9a). Using that approximate boundary condition, we find that

$$C_L = \frac{C_0}{1 + kL/D}, \tag{2.11a}$$

and, with Eq. (2.9b), the inner interface's rate of motion

$$\frac{dL}{dt} = \frac{\Omega k C_0}{1 + kL/D}. \tag{2.11b}$$

This last is a differential equation for $L(t)$ which is easily solved (the variables separate) under the initial condition $L(0) = 0$. We thus get the required thickness-time relation

$$L/k_L + L^2/k_P = t, \tag{2.12}$$

where the so-called *linear* and *parabolic* rate constants k_L and k_P are

$$k_L = \Omega k C_0 = \Omega k K p, \tag{2.13a}$$

$$k_P = 2\Omega D C_0 = 2\Omega D K p. \tag{2.13b}$$

We end this section with a

Discussion and Critique:
a. The oxidation law (2.12) is called *linear-parabolic* for obvious reasons. This kinetic behavior was recognized well before [7] Deal and Grove. The linear term on the left-hand side dominates for small enough oxidation times, while the quadratic term weighs more heavily for longer times. Most experimental observations confirm this behavior. In particular, the oxidation rate $\dot{L}(t)$ goes from a constant, for short times, to a $t^{-1/2}$ behavior, for long times. The experimental determination of the rate constants k_L and k_P is, however, no easy matter, because the transition from "pure" linear to quadratic behavior occurs over several orders of magnitude of time. The reader is encouraged to peruse Ref. [8], an experimental study using an automated ellipsometer, to understand just how difficult an unambiguous determination of these rate constants can be. These considerations are amplified in the following

Exercise 2.4: *Characteristics of the linear-parabolic law*
Equation (2.12) describes a parabola in the (t,L) plane. Solve for the function $L(t)$ and characterize this parabola (axis, branches, physical and nonphysical parts). Then expand that function in an appropriately chosen dimensionless time variable, and define times t_L and t_P such that linear kinetics hold for $t < t_L$, while parabolic kinetics dominate for $t > t_P$. The interval of time between t_L and t_P is called the region of *mixed* kinetics; it is also cause for much confusion in the literature. ∎

b. Both of Eqs. (2.11) involve the denominator $1+kL/D$. In the language of equivalent circuits (*cf.* Exercise 2.1), this can be understood as the sum of two resistances in series, k^{-1} and L/D, that contribute to the overall rate limitation of the oxidation process. For example, Eq. (2.11a) expresses the division of the total "potential drop" C_0 into a potential C_L across the surface resistance k^{-1}, and a potential C_0-C_L that develops across the diffusional resistance L/D. The dimensionless ratio kL/D measures the relative dominance of these two resistances. Such a decomposition into surface and bulk rate limitations is of fundamental importance in any rate process. There are many other kinetic processes of this type.

c. The measured rate constants k_L and k_P allow us to determine the primary constants of the model, k and D, *if* the solubility K is known precisely from independent measurements. If not, Eqs. (2.13) show that data-fitting oxidation experiments (at quasi-steady state) can only yield the *permeabilities* Kk and KD.

d. An example of the linear and parabolic rate constants' temperature dependence is given in Fig. 2.3, adapted from Ref. [9]. We note that they do *not* have a strict Arrhenius behavior $\propto \exp(-E/k_B T)$, where E is an activation energy, k_B is Boltzmann's constant $(8.62 \times 10^{-5}$ eV/K), and T is the absolute temperature. Therefore, we realize immediately that the simple model of Deal and Grove [4] cannot tell the whole story of silicon oxidation: The postulated kinetic steps cannot be *elementary steps* in the sense of chemical kinetics [10].

e. From Eqs. (2.11) we learn that both rate constants should be proportional to the oxygen partial pressure. While this has been properly assayed for the parabolic constant k_P, and thus gives credence to the model, the linear constant has a power dependence on p that appears to depend on temperature. In fact, some experiments indicate that $k_L \propto p^{\alpha(T)}$, where $\alpha \to 0.5$ for low temperatures ($\lesssim 750°$C) and $\alpha \to 1$ for high temperatures ($\gtrsim 1000°$C). The author has offered a model of oxidation that provides this dependence [11] and that is supported by further experimental evidence [12] regarding the curvatures in Fig. 2.3. Briefly, this model considers the possibility of O_2 dissociation at the Si/SiO$_2$ interface and the subsequent reaction of *atomic* oxygen with Si, in parallel, of course, with the reaction (2.7b). Again, one obtains linear-parabolic kinetics, with essentially the same long-time behavior as in the Deal-Grove model, but in which the linear rate

Fig. 2.3: Arrhenius plot of the linear and parabolic rate constants for silicon oxidation at one atmosphere of oxidant. Data from Irene and Dong [9, Fig. 2]; reprinted by permission of the publisher, *The Electrochemical Society, Inc.*

constant has the required non-integral and temperature-dependent "order of reaction." As mentioned earlier, a precise and unambiguous determination of oxidation rate constants is very difficult, and possible extensions of the Deal-Grove model must thus be regarded as tentative. Nevertheless, we are in possession of sufficient qualitative evidence to rule out certain processes. For example:

Exercise 2.5: *The impossibility of homogeneous reactions*
Work out the kinetics of SiO_2 growth if O_2 can react homogeneously with the glass network, and show that they are incompatible with the experimental evidence. [Hint: Include a sink term $-\kappa C$ in the diffusion equation; the other equations remain unchanged.] ∎

Exercise 2.6: *The impossibility of surface dissociation*
Assume that O_2 dissociates at the gas/SiO_2 interface and that *atomic* oxygen diffuses and combines with Si to form fresh SiO_2. Define the appropriate reactions and find the boundary conditions from mass balance. Show that this scheme is incompatible with the experimental evidence. ∎

These exercises indicate the nature of model building: It behooves the proposer of a given physico-chemical process to work out and discuss the consequences of any other imaginable scheme.

f. In its original formulation [4], the Deal-Grove model neglected, from the outset and without explanation, the term ΩC_L in Eq. (2.9a). Although this term is numerically small, its neglect in the boundary conditions is equivalent to a violation of mass conservation. The Deal-Grove model, as well as its formulation here, suffers from another logical inadequacy. It will be recalled that the origin of coordinates was chosen at the gas/SiO_2 interface (it could have been chosen at the other interface as well). Oxidation experiments, however, are conducted in a laboratory frame such that the substrate is at rest. In this frame *both* interfaces appear to move, and the diffusion equation for O_2 in SiO_2 must then contain convective terms that reflect these motions. These correction terms are likely to be negligible for oxidation processes, but they are certainly vital for the accurate description of impurity redistribution in oxides.

g. The model we have discussed is one-dimensional. The fabrication of device structures, however, presents problems [13,14] that cannot be understood without 2-*d* or 3-*d* modeling. Solutions are necessarily numerical [15], rather than analytic. In spite of its long history, the kinetics of SiO_2 formation does not form a closed field. It is still an object of active investigation [16].

h. Let us return and examine more closely the assumption of a quasi-steady state. From Eqs. (2.10), (2.11a), and (2.12) it is evident that the concentration distribution depends on time *through* the schedule $L(t)$, *i.e.*, $C = C(x,L(t))$. This situation should be familiar from the "adiabatic approximation" of mechanics. For example, the period of a pendulum *does* change with time if the pendulum slowly changes its altitude above sea level. This period, however, changes only through the change in the local gravitational acceleration. Similarly, many problems of phase change occur slowly enough that the fields (here, the concentration) adjust to their steady-state values corresponding to the instantaneous value of a forcing term (here, the oxide thickness). The rate of that forcing term (here, $\dot{L}$) is then determined by overall mass balance, as in Eq. (2.9b). The "goodness" of this approximation cannot be assessed in 1-*d*, as for this example, without detailed numerical work. The following sections, however, will provide an example (growth of spherical precipitates) for which there also exists an exact time-dependent solution. The desired comparison will be found in Chap. 5.

2.4 A FEW CONSIDERATIONS ON CAPILLARITY

The next example deals with nucleation and growth. Problems of this type require a clear understanding of a curved surface's effect on local thermodynamic variables. These variables are all fields in the sense of Sect. 1.8. In this section we review some *equilibrium* laws for small systems [17-19]. These laws follow from basic thermodynamics, which is summarized in Appendix B. The *kinetic* problem of nucleation and growth will be taken up in the next section.

Consider, for simplicity, a system of two phases, one of which might (but does not necessarily) contain the other, and each of which contains c components.† This system is represented schematically on Fig. 2.4. Assume that each phase can be completely described by its temperature T, pressure p, and chemical potentials μ_i ($i = 1,2,...,c$). This is equivalent to an assumption of liquid-like behavior, even though, later, we shall apply the results of this section to solids. These variables are not independent; their variations, *for whatever reason*, must satisfy the Gibbs-Duhem relation

$$Cs\, dT - dp + \sum_{i=1}^{c} C_i\, d\mu_i. \tag{2.14a}$$

For definiteness, the concentrations C_i are measured here in units of number density, and the total concentration is simply the sum over components

---

† Some of these components may be absent in one of the phases, in which case one must put their composition variables equal to zero. By *phase* one generally means a region of space throughout which *all fields are continuous*. These fields are, in general, continuously varying within a given phase, but at least one of them must suffer a discontinuity at phase boundaries.

$$C = \sum_{i=1}^{c} C_i. \tag{2.14b}$$

It follows that the entropy density s is defined per particle. Equations (2.14) must hold for each phase α ($\alpha = 1,2$). Therefore, *all* the above variables come "dressed" with an additional superscript to indicate which of the two phases they characterize. For example, $C_1^{(2)}$ is the concentration of the first component in the second phase.

Gibbs [17] showed long ago that the equilibrium of *each* phase requires the constancy of the T-, p-, and μ-fields over the domains $\mathscr{D}^{(1)}$ and $\mathscr{D}^{(2)}$. For example, $\nabla p^{(2)} = 0$ in the second phase if it is at equilibrium.[†] Moreover, equilibrium *between* phases demands the following equalities at the phase boundary Σ:

$$T^{(1)} = T^{(2)}, \tag{2.15a}$$

$$p^{(1)} - p^{(2)} = \gamma K, \tag{2.15b}$$

$$\mu_i^{(1)} = \mu_i^{(2)} \quad (i = 1,2,...,c). \tag{2.15c}$$

We see that the temperature and chemical potentials are continuous across the phase boundary, but that the jump in pressure is proportional to the surface tension γ and to the local curvature K. This Eq. (2.15b) is called Laplace's law of capillarity, but it was certainly known earlier to Young. The "curvature" in that equation is the *mean curvature* of a surface as defined in differential geometry [20], *i.e.*, the sum of inverse radii of curvature of curves formed by the intersection of the surface with the two "principal" planes containing the unit normal n.[‡] Referring to Fig. 2.4, the unit normal is directed from phase 1 to phase 2, and the mean curvature is reckoned positive if the surface is convex towards the second phase. For example, $K = 2/R$ or $1/R$ for a sphere or cylinder of radius R.

Equations (2.14-15) are all that is required to describe the "equilibrium of heterogenous substances," in general. To that end, it is

---

[†] This statement is strictly true only for isotropic phases at rest and in the absence of external fields [17,19]. This same remark holds for Laplace's law (2.15b).

[‡] The mean curvature obeys an invariance property known as Bonnet's theorem: K can be computed from *any* two orthogonal sections of a surface.

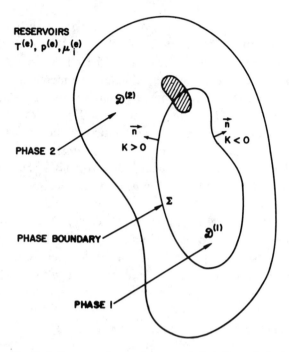

Fig. 2.4: System of two phases and their boundaries.

convenient to imagine our two-phase system immersed in a large external phase with which it can reversibly exchange heat at temperature $T^{(e)}$, mechanical work at pressure $p^{(e)}$, and masses at chemical potentials $\mu_i^{(e)}$. This idealization requires comment. The external phase represents interactions with the world at large, *i.e.*, the experimental conditions that can be imposed. This phase is also assumed large; therefore, changes in the enclosed phases, 1 and 2, induce negligible changes within it, and surface effects at its boundaries play a negligible role. Such a phase is often called a system of "reservoirs."

We now consider a special case: a two-component system at constant external temperature $T^{(e)}$ and pressure $p^{(e)}$. We then seek the chemical potentials' dependence on the curvature K.† Call these two components A and B. The situation is now very simple if we mentally travel inward, so to speak, from the external phase towards phase 1. First, constant experimentally imposed $T^{(e)}$ implies constant $T^{(2)} = T^{(e)}$ at the second phase's external boundary, because a relation of the type (2.15a) must hold there. Since the temperature $T^{(2)}$ is a constant field in phase 2, its value $T^{(e)}$ "propagates," unchanged, throughout that phase. Likewise, $T^{(1)} = T^{(e)}$ by the same continuity arguments. Therefore, we have $dT^{(e)} = dT^{(1)} = dT^{(2)} = 0$. Then $dp^{(e)} = dp^{(2)} = 0$ for similar reasons. Note, however, that $dp^{(1)} = d(\gamma K) \neq 0$ because of Eq. (2.15b). In other words, the phase boundary Σ is a barrier to the propagation of $p^{(e)}$ into phase 1. The four chemical potentials, on the other hand, cannot all be maintained constant without violating the phase rule (see the previous footnote). Nonetheless, they are constant fields over their respective domains. By virtue of Eqs. (2.15c), their values, for each component, must be equal throughout the two-phase system. If μ_A and μ_B are these values common to both phases, and if $d\mu_A$ and $d\mu_B$ are their variations, then the above considerations show that Eqs. (2.14-15) reduce to

$$-d(\gamma K) + C_A^{(1)} d\mu_A + C_B^{(1)} d\mu_B = 0, \qquad (2.16a)$$

$$C_A^{(2)} d\mu_A + C_B^{(2)} d\mu_B = 0, \qquad (2.16b)$$

from which, for example, we can eliminate $d\mu_A$:

$$-d(\gamma K) + \left(C_B^{(1)} - C_A^{(1)} C_B^{(2)} / C_A^{(2)} \right) d\mu_B = 0. \qquad (2.16c)$$

If, now, phase 2 is a dilute solution of component B, then the second term in parenthesis must be small. If, further, the concentration of that

---

† The thermodynamic variables that can be held constant and those whose variations we seek must be compatible with the *phase rule*; see Appendix B. According to Eq. (B.6) of that appendix, here we have three degrees of freedom. From the nine-member list (B.5) we choose fixed values of $T^{(2)}$ and $p^{(2)}$ by maintaining the heat and work reservoirs at the constant values $T^{(e)}$ and $p^{(e)}$. We then inquire, for example, into the functional dependence of the chemical potentials on the mean curvature K. The external values $\mu_A^{(e)}$ and $\mu_B^{(e)}$ must be left free to adjust.

component in phase 1 does not vary much with the curvature of that phase, then Eq. (2.16c) can be integrated to give

$$\mu_B(K) - \mu_B(0) = \gamma K / C_B^{(1)}. \tag{2.17a}$$

It must be understood that the integration is carried out from the state for which the surface is flat ($K = 0$, which means that phase 1 is effectively of infinite extent) to the state for which the curvature is finite. Moreover, that integration requires no assumption whatsoever regarding the surface tension γ's dependence on curvature or composition.

Chemical potentials are fine objects, but one tends to measure concentrations instead. We can easily find the equilibrium concentration's dependence on curvature from a known $\mu(C)$ relation in the second phase. In a solution (solid, liquid, or gas), we know that [21]

$$\mu_B^{(2)} = \mu_B^0(p^{(2)}, T^{(2)}) + kT^{(2)} \ln a_B^{(2)}, \tag{2.17b}$$

where μ_B^0 is the chemical potential of pure component B, $a_B^{(2)}$ is the activity of the minor component B in solution, and k is Boltzmann's constant. Inserting Eq. (2.17b) into (2.17a), we find an exponential dependence of activity on curvature. In particular, for very dilute solutions of component B, the activity tends towards a multiple of that component's mole fraction, and we get the final result

$$C_B^{(2)}(K) = C_B^{(2)}(0) e^{\gamma K / kTC_B^{(1)}}, \tag{2.17c}$$

attributed to Ostwald.† The quantity $\gamma/kTC_B^{(1)}$ is positive and has dimensions of length. It is called the *capillary length* Γ. Therefore, the formula (2.17c) shows that the equilibrium concentration of the minor component above a convex surface element ($K > 0$) is larger than what would prevail over a flat element, and *vice versa* for concave surface elements.

The application of the formula (2.17c) to equilibrium phase diagrams is of the utmost importance. Figure 2.5 shows a simple diagram that corresponds to the equilibrium of a solid compound AB (phase 1) with

---

† Results of this type — shifts in equilibria due to curvature — are generically known as "Gibbs-Thomson effects." For example, the effect of curvature on vapor pressure is attributed to Lord Kelvin, who, after all, was also W. Thomson. For convenience, some of these results are collected in Appendix B.

a liquid solution of B in A (phase 2). A concrete example might be the crystallization of GaAs from a liquid solution of As in Ga. Here, the solidus is essentially a vertical line at the stoichiometric AB composition, while the liquidus is curved and depends exponentially on temperature. The calculation of this liquidus, for GaAs, assumes that the solution is regular and that both phases are infinite in extent [21,22]. This phase diagram tells us that, at a given temperature (and pressure), a *unique pair* of solid and liquid compositions is in equilibrium. The one-to-one corresponding points are labeled (1) and (2) in Fig. 2.5.†
The region above the liquidus, the so-called *single-phase region*, represents thermodynamic states of the liquid that are not in equilibrium with any finite solid. We are then free to choose arbitrarily the solution's temperature and concentration, as is evident to any good cook. On the other hand, the region of Fig. 2.5 bounded by the solidus and liquidus lines represents metastable states of the liquid; it is called the *two-phase region* for the following reason. Assume that the solid compound is of small extent, and therefore that its curvature is, on the

Fig. 2.5: Schematic phase diagram for a binary liquid-solid system. The capillarity-induced liquidus shift is shown dashed.

---

† This is compatible with the phase rule (B.4) of Appendix B for large systems. Because $f = 2$, we can arbitrarily choose experimental conditions of constant T and p. Then, however, all other quantities, such as equilibrium compositions, are fully determined.

whole, large and positive. It follows from Eq. (2.17c) that this solid would be in equilibrium with a liquid whose concentration $C_B^{(2)}(K)$ in the minor component B is *larger* than it would be for an infinite solid. This is true at any temperature below the compound's temperature of fusion T_F. Thus, a given positive curvature K of the solid phase 1 causes a *liquidus shift* (dashed) to the right. Reciprocally, given an arbitrary point of the two-phase region [labeled (K) in Fig. 2.5], then there surely exists a solid, and one only, small enough (*i.e.*, of positive and large enough curvature K) that is in equilibrium with the liquid of composition $C_B^{(2)}(K)$.† In the next section we shall see that these small solid particles are metastable, *i.e.*, there exists a critical dimension of these particles, related to the capillary length, below which the solid dissolves and above which it grows.

Exercise 2.7: Justification of Ostwald's formula

Carefully justify all the steps leading from Eq. (2.16c) to (2.17c). [Hint: Recall that the mole fraction of component B, in either phase, is defined as $x_B = C_B/C$, $C = C_A+C_B$; therefore $x_A+x_B = 1$. For example, take an activity of the form $a = \alpha x+\beta x^2$.] What can you say if the solution is *not* dilute? Consider, first, the case of a stoichiometric AB compound, then a pure solid phase 1 composed of B only. In all cases, discuss also the $C_A(K)$ dependence of the major component. ∎

Exercise 2.8: Application to GaAs

Take the case of GaAs, a prototypical binary III-V system. Compute the capillary length at $700°C$ and $1000°C$, knowing that $\gamma \cong 860$ erg/cm^2, that solid gallium arsenide's density is 5.2 g/cm^3, and that its molecular mass is 72.3 g/mole. With these values and Eq. (2.17c), compute the corresponding enhancement of As in solution for 0.01- and 1-μm-diameter spheres. ∎

Finally, we note that the derivation of Ostwald's formula (2.17c) was predicated on the implicit assumption that phases 1 and 2 are in *global* equilibrium. This is not necessary: Consider an *arbitrary* domain (shown hatched in Fig. 2.4) that "straddles" both bulk phases and their phase boundary. If that system is in thermodynamic equilibrium,

---

† All these considerations are compatible with the phase rule (B.6) for small systems.

then one can show that Eqs. (2.14-15) hold *locally* because the domain is arbitrary. Consequently, Eq. (2.17c) also holds locally. This remark is particularly useful when considering kinetic processes, as we shall shortly. Although a system of two bulk phases and a phase boundary can be far from equilibrium — and we then wish to compute its rate of relaxation towards equilibrium — it is often useful to assume that the phase boundary, itself, is locally at equilibrium with the adjacent bulk phases. In addition, the formula (2.17c) applies regardless of the mean curvature's sign. A corrugated solid body has surface elements with negative as well as positive curvature (see also the next footnote). If an element has a *negative* mean curvature K', then $C_e(K')$ is the concentration in the liquid phase which is in equilibrium with that element. This causes a local liquidus shift to the *left* on Fig. 2.5. These considerations are key to the understanding of morphological features on surfaces.

2.5 THE PRECIPITATION OF SPHERICAL PARTICLES

Problems of nucleation and growth occur throughout our experience. They are also crucial to our ability to manufacture parts having required mechanical and electrical properties. Whether rain precipitates from clouds, or carbon precipitates from steel, depends, in large measure, on the surface properties of small phases. These phases can exchange matter and energy with a large phase, the *matrix* or *parent* phase. The small phases are called *nuclei* because, as they grow in extent at the expense of the matrix, they are the cause of larger, macroscopically observable phases. The origin, shape, and thermodynamic properties of nuclei are still open to conjecture, although molecular-dynamics computer simulations [23] do shed some light on this problem. In this section, we solve a problem that is again mathematically very simple, but that will provide us with a clear understanding — and a dynamic one, at that — of the critical radius of nucleation.

Consider again a parent phase 2, in any state of aggregation (gas, liquid, or solid), that is a *homogeneous* mixture of two chemically distinct species A and B. We then ask if this mixture, at a given temperature, pressure, and composition, can promote the formation of another phase 1, the nucleated *precipitate*, containing these same species. In

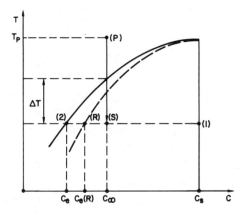

Fig. 2.6: Same as Fig. 2.5 in simplified notation. The undercooling ΔT is equivalent to the supersaturation measures σ or σ_r.

other words, can the initially homogeneous mixture become heterogeneous, and if so, at what rate? Without inquiring into the origin of nuclei, it is clear that, once formed, they will grow (or decay) because of mass or heat transport. And transport implies diffusion over macroscopic distances.

Figure 2.6 is a phase diagram similar to Fig. 2.5. Again, the diagram is drawn for a system of liquid parent phase and solid precipitate, but the considerations that follow are independent of the system's state of aggregation. We also simplify the the notation of the last section by putting $C_s = C_B^{(1)}$ and $C_e = C_B^{(2)}(0)$ for the equilibrium concentration of B in a precipitate and the parent phase, respectively. At given temperature and pressure, we know that the compositions C_e and C_s are corresponding equilibrium compositions if both phases are large in extent. Consider, now, a *spherical* precipitate of radius R immersed in the parent phase. We also know that there exists, under the same conditions of temperature and pressure, a certain composition $C_e(R)$ of the parent phase that would be in equilibrium with this (small) precipitate. That composition is given by Eq. (2.17c), or

$$C_e(R) = C_e\,e^{2\gamma/RkTC_s} \tag{2.18}$$

in the present notation. It must necessarily lie in the two-phase region of the diagram because the sphere has a positive (and constant) mean curvature $2/R$. Therefore, we have the inequalities

$$C_e < C_e(R) < C_s.$$
(2.19)†

The dynamic situation we wish to examine is as follows. Imagine that we prepare, at a given temperature T, a homogeneous solution of constant composition C_∞. Its state (S) can lie anywhere in the "half strip" $0 < C < C_s$ and $T > 0$ of the phase diagram in Fig. 2.6; here, it is drawn in the two-phase region. Assume now that the Lord shows his hand through the creation, somewhere in the bulk of this solution, of a solid spherical nucleus of radius R_0. Assume, also, that the surface of this nucleated precipitate is active enough for exchanges of mass to occur rapidly with the solution. The concentration in solution C_R, right next to the precipitate's surface, is then the equilibrium concentration $C_e(R)$ corresponding to its instantaneous radius.‡ Since, in general, $C_e(R) \neq C_\infty$ (remember that the solution's overall composition was arbitrary), it follows that diffusion must occur, the local concentration C of component B in solution must vary, and, thus, the precipitate must grow or dissolve.

The mathematical problem, in its full generality, is quite complex because we have, here again, a problem of phase change with a moving boundary. The quasi-steady-state approximation, however, leads to a simple solution, much as for silicon oxidation. Figure 2.7 shows the concentration profile that corresponds to the conditions on Fig. 2.6. [Diffusion *in* the solid precipitate is often negligible.] In the absence of other information we assume spherical symmetry around the precipitate's center, which is also the origin of our coordinate system. In that system, the precipitate's boundary moves according to a schedule $r = R(t)$ that we must find. Towards that end, recall the diffusion equation (1.36) for the minor component B in solution,

$$\frac{\partial C}{\partial t} = \frac{D}{r^2}\frac{\partial}{\partial r}\left(r^2\frac{\partial C}{\partial r}\right),$$
(2.20)

---

† In passing, we note that the impossibility of finding a nucleus in the single-phase region (the region of good cooks, above the liquidus), and thus $C_e(K) < C_e$, is related to the impossibility of finding *finite* bodies of constant negative mean curvature.

‡ This is equivalent to the assumption of *local equilibrium* that was mentioned at the end of the last section.

Fig. 2.7: Schematic concentration distribution for the problem of spherical precipitate growth.

written in spherically symmetric coordinates.† We must find its solution $C(r,t)$ subject to the boundary conditions

$$C(\infty,t) = C_{\infty}, \tag{2.21a}$$

$$C(R,t) \equiv C_R = C_e(R), \tag{2.21b}$$

and an appropriate initial condition that need not concern us here. The first condition, sometimes called a "far-field" condition, simply indicates that the concentration in the solution must approach its prepared value at points far enough from the precipitate. The second condition is our real concern. In accordance with an earlier assumption, it simply states that the concentration in solution C_R at the precipitate's surface must equal its equilibrium value, *compatible* with the sphere's instantaneous radius. That value is given by Eq. (2.18), which is nonlinear in the radius R. Moreover, that radius changes with time in a yet unknown way.

---

† The following calculations are valid only when convection in the liquid is negligible. In particular, note that convection can be caused by density changes in the liquid-solid transition (see Ref. [24] for a complete account of these effects).

Any moving-boundary problem still requires a Stefan condition for the precipitate's boundary motion; global mass conservation, again, provides the "royal way." Consider the total mass of component B

$$M(t) = \int_0^{R(t)} C_s \, dV + \int_{R(t)}^{\infty} C(r,t) \, dV,$$

where the volume element is $dV = 4\pi r^2 dr$. The calculation is now very simple because there is no reason for M to change.† With the condition $\dot{M} = 0$ and the use of Leibnitz's rule and Eq. (2.20), we get‡

$$D \frac{\partial C}{\partial r} \Big|_{r=R} = \dot{R}(C_s - C_R), \tag{2.21c}$$

with an evident geometric interpretation: Any increase of the precipitate's total mass must be strictly accounted for by a depletion of the component B in the parent phase. In other words, the volumes corresponding to the two hatched areas of Fig. 2.7 must be equal.

The solution of this problem is very simple in the quasi-steady-state approximation. Assume, then, that the left-hand side of Eq. (2.20) is small; its right-hand side reduces, therefore, to an ordinary differential equation whose general solution is

$$C(r,R(t)) = C_{\infty} - (C_{\infty} - C_R)R/r, \tag{2.22}$$

in accordance with the general result (2.4b). Here, we anticipate that the concentration's time dependence will occur again through the boundary's schedule, as in our earlier oxidation problem. This schedule is easily determined by inserting Eq. (2.22) into the Stefan condition (2.21c). Thus

$$\frac{dR}{dt} = \frac{D}{R(C_s - C_R)}(C_{\infty} - C_R). \tag{2.23a}$$

Finally, the as yet unknown surface concentration C_R must be compatible with the equilibrium condition (2.18), and we get the differential equation

---

† Recall that we assume all exchanges of matter at the precipitate's surface to occur at equilibrium (*cf.* Exercise 2.11 below).

‡ See the footnote following Eq. (1.20). We assume that the flux at infinity decreases faster than r^{-2}; this is proved in Appendix C.

$$\frac{R}{D}\frac{dR}{dt} = \frac{C_\infty - C_e\, e^{2\Gamma/R}}{C_s - C_e\, e^{2\Gamma/R}},$$ (2.23b)

where it is again convenient to introduce the capillary length $\Gamma = \gamma/kTC_s$. The problem is, at least formally, solved: The solution of the *nonlinear* differential equation (2.23b), with the initial condition $R(0) = R_0$, yields C_R because of Eq. (2.18), and then the full concentration distribution C follows from Eq. (2.22).

Let us look at the qualitative features of the differential equation (2.23b). It gives the precipitate's rate of growth in terms of purely thermodynamic variables on its right-hand side. Observe first that the denominator of that right-hand side is always positive because of the inequalities (2.19). The numerator, however, can change sign, and the precipitate's size for which this occurs is called the *critical radius of nucleation R^**. Thus, $C_e(R^*) = C_\infty$ implies

$$R^* = \frac{2\Gamma}{\ln(C_\infty/C_e)} = \frac{2\gamma}{kTC_s \ln(C_\infty/C_e)},$$ (2.24)

and this radius corresponds to a *turning point*† of the differential equation: The precipitate's rate of growth vanishes when $R = R^*$. In addition, inspection of Eq. (2.23b) shows that if $R < R^*$ then $\dot{R} < 0$, and the precipitate *dissolves*. Likewise, *growth* obtains if $R > R^*$. Thus, although the precipitate is in equilibrium with a solution of given temperature and composition when its radius R is exactly equal to the critical radius R^*, that equilibrium is *metastable*: any deviation from the equilibrium condition produces further changes that drive the system farther away from equilibrium.‡

Another way of expressing this same result arises if we observe from Eqs. (2.23) that $\dot{R}$ has the sign of $C_\infty - C_e(R)$. Thus, if the parent phase's bulk state (S) lies to the right of state (R), as in Fig. 2.6, then the precipitate will grow. In the language of thermodynamics, the

---

† The situation should be familiar from mechanics. For example, the position(s) of a mass point at which its velocity vanishes and, therefore, at which the motion reverses itself are the turning points of its equations of motion. These are second order in time, in contrast to the first order equation above.

‡ In contrast to the usual equilibrium arguments for R^*, we emphasize that here its calculation is dynamical.

uid is *supersaturated* with respect to a solid of radius R. This is true or *all* radii larger than the critical radius R^*. Therefore, the prepared state (S) of the liquid is supersaturated with respect to all states (R) lying to its left. The converse statements regarding states to the right of (S) then define the concept of *undersaturation* in a natural way. Of course, if (S) lies to the left of the liquidus itself, then the liquid will always be undersaturated with respect to solids, no matter what their (positive) radii of curvature might be. In that context, one sometimes finds these notions expressed equivalently in terms of temperature differences, and these are useful experimentally. The supersaturated state (S), shown in Fig. 2.6, can be achieved by carefully cooling a solution, of prepared composition C_∞, from a temperature T_P above the liquidus, down through the liquidus to the final state (S). The maximum amount of available supersaturation $C_\infty - C_e$ is equivalent to the *undercooling* ΔT shown on the diagram. Both are measures of the maximum driving force of the precipitation process.

Exercise 2.9: Undercooling and critical radius

Discuss a prepared solution's ability to withstand a given undercooling. Assume that $C_e = C_0 e^{-\Delta H/kT}$ and $\partial \gamma / \partial T \leq 0$. In particular, how does R^* change with temperature? Express R^* also in terms of a chemical-potential difference. ∎

We now briefly turn to the integration of Eqs. (2.23). First, we see that capillarity plays a meager role when R is large enough. A precipitate, in fact, always grows if a given nucleus has an initial radius $R_0 > R^*$. If, in addition $R_0 \gg 2\Gamma$, then $C_e(R) \cong C_e$ for all subsequent radii. Then, it follows that Eq. (2.23b) reduces to

$$\frac{R}{D} \frac{dR}{dt} = \frac{C_\infty - C_e}{C_s - C_e}. \qquad (2.25a)$$

The right-hand side is constant and is called the *absolute* supersaturation σ, reserving the word *relative* supersaturation for the ratio $\sigma_r \equiv \ln(C_\infty / C_e)$ that occurs in the definition of the critical radius (2.24). The solution of Eq. (2.25a) is evidently the parabolic law

$$R(t) = (R_0^2 + 2\sigma Dt)^{1/2}, \qquad (2.25b)$$

and it represents the asymptotic behavior of a precipitate's increase in size if it had the good fortune to be "launched" with a large enough radius. It is interesting to note that both relations (2.25) hold *exactly* if the surface tension γ vanishes. If the previous conditions are not met, then one must integrate the nonlinear equation (2.23b). This we will not do, because such calculations would carry us far afield into the kinetics of precipitation. To compute these properly would still require an estimate of the *rate* at which nuclei are formed and estimates of their initial size distribution. And these are *not* diffusion problems.

Exercise 2.10: *Supersaturation measures and length scales*
Discuss the physical meaning of both measures of supersaturation σ and σ_r. In particular, what is σ_r if it is "small"? Note that there are *two* characteristic length scales for the rate equation (2.23b): R^* and 2Γ. The first measures deviations from metastability; the second measures finite-size effects. Discuss that equation for all possible initial radii R_0 relative to these lengths. Then discuss the solution (2.25b) for both positive and negative σ. ∎

Exercise 2.11: *Effect of finite surface kinetics*
Assume that exchanges at the precipitate's surface with the adjacent bulk phases occur at a finite rate. Specifically, assume that the solid-liquid phase transformation can be described by a "reaction rate" $k[C_R - C_e(R)]$, where k is a "reaction constant." Discuss its meaning, and find the appropriate boundary and Stefan conditions. [Hint: Consider mass balance in each phase.] Next, find the rate equation which generalizes Eq. (2.23b). What is then the critical radius of nucleation? ∎

There are many other questions relating to the growth of precipitates. For example, we assumed a spherical shape that was preserved during growth or dissolution. Is this possible? And is the sphere the only "shape preserving" configuration? These questions are generally difficult to answer in full. We refer to the paper by Ham [25] for a partial answer. In brief, there exist a class of "quadric" surfaces that maintain their shape under diffusion if capillarity effects are neglected. The *stability* of these surfaces, on the other hand, rests crucially on such effects. The "morphological stability" of growing crystals was first studied in the early 1960's by Mullins and Sekerka (see Ref. [26] for an excellent review), and since then these questions have been actively pursued in the larger context of "pattern formation" [27]. Indeed, one may ask how a snowflake gets its orderly shape, even as it is born out of an unordered and essentially homogeneous phase. The scientific and technological importance of these questions cannot be underestimated.

Exercise 2.12: *Precipitation of cylindrical particles*
Discuss equilibrium over cylindrical surfaces and find the analog of Eq. (2.18). Compute the kinetics of long cylindrical precipitates and estimate the critical radius of nucleation. ∎

Another problem that deserves mention is "Ostwald ripening." We assumed, in effect, that there were few nucleation events in the parent phase. If, however, many nuclei are produced, then the diffusion fields around each precipitate can overlap and compete for the available amount of supersaturation. Big precipitates grow at the expense of smaller ones, and the bulk composition C_∞ decreases with time as the population of precipitates changes. One says that the solution "ages" [28]. The calculation of this time dependence and the precipitates' ultimate size distribution is the aim of a theory proposed by Lifshitz and Slyozov [29,30].

Finally, this section has dealt only with problems of *homogeneous* precipitation, *i.e.*, when nucleation occurs in the interior of a parent phase. This must be contrasted with *heterogeneous* precipitation, a subject in its own right. Then, nucleation events and precipitate growth *at* phase boundaries are the main concern [31,32].

2.6 CRYSTAL GROWTH ACCORDING TO THE THEORY OF BURTON, CABRERA, AND FRANK

The years 1949-1951 witnessed a truly remarkable development in our understanding of why crystals grow at all [33]. Indeed, a dilemma arises if we consider the structure of crystal surfaces. To grasp this question, observe that there are essentially two types of crystal surfaces, those whose orientation lies or does not lie near a close-packed direction. The former surfaces, called *vicinal*, are mostly smooth on an

Fig. 2.8: Schematic TLK model of vicinal surfaces.

atomic scale. They consist of close-packed surface elements called *terraces* that are separated by risers called *steps*. These steps are necessary to produce a surface with a given average orientation. The steps themselves are not simple structures, because of thermal fluctuations. Indeed, it had long been known [34] that they roughen to produce re-entrant corners called *kinks*. Figure 2.8 shows how a vicinal surface is composed of terraces and steps, and how steps are composed of ledges and kinks. With simple bond models one can show that kinks are privileged surface sites: Adding or subtracting molecules at these sites cannot change the total cohesive energy of a large crystal. Consequently, the binding energy at a kink site must be related to the equilibrium chemical potential. From a dynamic point of view, molecules from the parent phase can attach to terraces and diffuse towards ledges. Once attached to ledges, they diffuse towards kinks where they find their ultimate resting place. In that process, the kink position moves in the negative y-direction of Fig. 2.8, and when that kink reaches the crystal's edge, the whole step will have progressed by one unit in the positive x-direction. In turn, when a step moves towards a crystal edge, then the crystal as a whole will have grown by one unit in the positive z-direction. This model is therefore called the "terrace-ledge-kink" (TLK) model of crystal growth; it was formulated long ago by Kossel, Stranski and Volmer [35]. In this sense, kinks are the ultimate agents of growth. More generally, they are privileged reaction sites on crystal surfaces. But, if kinks arise from fluctuations in steps, then what is the origin of steps? First, it can be shown that steps *cannot* be produced by thermal fluctuations, at least, below a critical temperature that characterizes two-dimensional melting [33]. Second, if these steps were due to intentional misorientation, then their advance to the crystal's edge would cause their annihilation, and, eventually, the growth process would stop. Third — and here is the dilemma — although steps can be produced by the edges of disk-like nuclei on terraces, it can also be shown that the supersaturation required for such heterogenous nucleation is often much larger than is observed experimentally. Before addressing this question, let us remark that crystal surfaces whose average orientation is far from close-packed are necessarily *rough* on an atomic scale. Each surface element contains many kink sites, but steps and terraces are no longer identifiable. Each of these kinks is a nucleation site, and the energy required for nucleation is far lower than on close-packed surface elements. There is then no appreciable barrier to crystal growth. For example, the surface of the spherical precipitate of the last section was implicitly assumed to be rough.

Thus, we rephrase our question: What is the origin of steps on vicinal surfaces? In 1949, F.C. Frank, one of the founders of modern dislocation theory, suggested that a screw dislocation (more precisely, a dislocation with a screw component) that intersects a crystal surface can provide an inexhaustible source of steps. This built-in imperfection bypasses the need to overcome a heterogenous nucleation barrier. The structure of steps, their motion, and the connection with the growth of crystals was then explored in a now classic paper by Burton, Cabrera, and Frank [33]. This theory is known as the BCF model of crystal growth, and it ranks high among twentieth-century achievements in physics. It must be remembered that these authors postulated a mechanism that was beyond the experimental observations of the time, but that has been amply confirmed since then.

A complete exposition of the BCF theory is beyond our scope. It does provide, however, a very interesting problem in diffusion theory. Consider a single step produced by an emerging screw dislocation. This step reaches out from the dislocation core to an edge of the crystal, and one shows that it is straight, on average, if the crystal is in equilibrium with the parent phase. Assume, now, that the supersaturation in the parent phase is "switched on." Molecules from that phase are then transferred by 3-*d* diffusion to the terraces; on terraces these adsorbed molecules execute 2-*d* jumps towards ledges, and, once attached to ledges, the molecules diffuse one-dimensionally towards the 0-*d* sites that are kinks. We thus have a nested set of diffusion problems that are distinguished by their dimensionality.† As explained earlier in connection with the TLK picture, the dislocation step will move. But it is also pinned at its core, which means that regions of the step far from the core will move at a velocity controlled essentially by the supersaturation, while those parts near the core will move less. Therefore, the dislocation step must wind itself around the core to produce a spiral. The dynamics of this spiral were explored in depth by BCF. Here, suffice it to say that its shape is essentially determined by the critical radius R^* of heterogeneous nucleation, which determines the minimum radius of curvature that prevails at the dislocation core. In effect, as with homogeneous nucleation, if any element of the dislocation step had a radius of curvature *less* than R^*, then it would recede rather than advance. The spiral, in polar coordinates (r, φ), was shown by BCF to be essentially archimedean

$$r \propto R^* \varphi. \tag{2.26a}$$

Therefore, the distance ℓ between successive arms of the spiral is constant and is proportional to R^*. Since the critical radius for heterogeneous nucleation is also inversely proportional to the relative supersaturation $\sigma_r = \ln(C_\infty / C_e)$ [see Eq. (2.24)], we get the important relation

$$\ell \propto \sigma_r^{-1} \tag{2.26b}$$

that relates step distance to supersaturation. The constant of proportionality in this equation depends on such things as the edge free energy of a 2-d nucleus, the temperature, and the interaction between competing spirals.

The full diffusion problem outlined above is evidently complex. It is good to remember, however, what we really want, namely an expression for the rate of growth V of a crystal surface. Figure 2.9 shows a normal section of this surface and two of its equidistant steps. The growth rate normal to a given close-packed orientation is equal to the velocity v of a step times the tangent of the misorientation angle ϑ. If h denotes the step height, then $\tan \vartheta = h/\ell$. Further, if J_s denotes the

---

† It can also be shown that "direct" processes, such as parent-phase→kink exchanges, are unlikely.

surface flux along terraces, then mass-balance considerations dictate that $vh/\Omega = 2J_s|_{step}$, where Ω is the volume of a molecule. The factor 2 comes from the two fluxes on either side of a given step. Thus we get an expression for the growth rate

$$V = (2\Omega/l)J_s\big|_{step}, \tag{2.27}$$

and the whole calculation reduces to the computation of molecular fluxes at steps.

Following BCF, we now make the simplifying assumptions:

a. Diffusion in the parent phase and along ledges is fast enough not to be rate limiting. The problem is thus reduced to the consideration of transport processes on terraces.

b. The kinks on a given step are so close together that, in view of the assumption of fast diffusion on ledges, steps are effectively line sinks.

c. Exchanges at kinks are effective enough to maintain the concentration at steps close to its equilibrium value.

d. The steps are so far from the dislocation core that they can be considered straight and equidistant. This implies a 1-d diffusion field in the x-direction on terraces.

e. The motion of steps is slow enough, compared to mass redistribution, that the quasi-steady state obtains, and Eq. (2.27) holds.

With these assumptions, the concentration $C_s(x,t)$ of adsorbed molecules on a given terrace must obey

$$\frac{\partial C_s}{\partial t} \cong 0 = -\frac{\partial J_s}{\partial x} + J, \tag{2.28a}$$

which is nothing but a balance law (1.47) (on a surface) with a "source" equal to the net flux J of molecules exchanged with the bulk parent phase. The flux J is then expressed as the detailed balance of an adsorption rate KC_∞ and a desorption rate C_s/τ_s, where K is the adsorption rate constant, C_∞ is the (constant) bulk concentration, and τ_s is the lifetime or residence time of an adsorbed molecule on the terrace. Thus, we have

$$J = KC_\infty - C_s/\tau_s. \tag{2.28b}$$

Fig. 2.9: Section through a vicinal surface. Note the adsorption, desorption, and diffusive processes on the surface.

It remains only to express the surface flux according to a phenomenological relation

$$J_s = -D_s \frac{\partial C_s}{\partial x}, \tag{2.28c}$$

which defines the surface diffusivity D_s.†

Exercise 2.13: *Balance laws for vicinal surfaces*
Choosing appropriate domains on the surface, prove that mass conservation implies Eq. (2.28a), and find the general Stefan condition for steps that leads to Eq. (2.27). ∎

The problem is now very simple. Observe, first, that we can relate the equilibrium concentrations in the bulk and surface phases because $J = 0$ at equilibrium. With Eq. (2.28b) we then get a "mass-action" relation

$$C_{s,e} = \tau_s K C_e, \tag{2.29a}$$

i.e., the equilibrium surface concentration in terms of the equilibrium bulk value. Then, we noted in Exercise 2.10 that C_∞ is given approximately by

---

† Note that the dimensions of the various quantities are $[C_s] = \text{cm}^{-2}$, $[J_s] = \text{cm}^{-1}\text{s}^{-1}$, but that D_s has the dimensions (cm^2/s) characteristic of all diffusivities [*cf.* remark (d) of Sect. 1.2].

$$C_\infty \cong C_e(1 + \sigma_r) \tag{2.29b}$$

if the relative supersaturation is small. Inserting Eqs. (2.28b,c) and (2.29) into Eq. (2.28a), we get the differential equation

$$D_s \tau_s \frac{d^2 C_s}{dx^2} - C_s + C_{s,e}(1 + \sigma_r) = 0, \tag{2.30a}$$

in which the supersaturation is evidently a forcing term. This equation also shows that the combination of physical quantities

$$x_s = (D_s \tau_s)^{1/2} \tag{2.30b}$$

must be a fundamental length scale. It is called the *mean diffusion distance*, and it plays the same role, here, as does the mean free path in other kinetic theories. Now, what about boundary conditions? By virtue of assumption (c), the concentration at steps must equal its equilibrium value. Since the balance equation (2.30a) holds on terraces between any two adjacent steps, and because this equation is symmetric under reversals of space, it behooves us to choose a coordinate system in Fig. 2.9 whose origin is placed symmetrically, right between adjacent steps. The boundary conditions then read

$$C_s(\pm \ell/2) = C_{s,e}. \tag{2.30c}$$

Equation (2.30a) is a second-order differential equation with constant coefficients; by standard methods, its solution is

$$\frac{C_s(x) - C_{s,e}}{\sigma_r C_{s,e}} = 1 - \frac{\cosh x/x_s}{\cosh \ell/2x_s}. \tag{2.31a}$$

Hence, we compute the flux (2.28c) and then the growth rate (2.27)

$$V = (C_{s,e}\sigma_r \Omega/\tau_s)(2x_s/\ell) \tanh(\ell/2x_s). \tag{2.31b}$$

The growth rate is thus the product of three terms. The first, $C_{s,e}\sigma_r\Omega/\tau_s$, is due to the net flux from the parent phase; it is the maximum possible growth rate. In fact, when $\ell \ll 2x_s$ the two other terms go to unity because the whole surface, effectively, becomes a sink.† The second term, $2x_s/\ell$, shows that only those molecules that impinge in a band of width $2x_s$ (the "catchment" area or capture cross-section) about the step are effective for growth. Molecules adsorbed at a distance greater than x_s from a step will have a large probability of desorbing before they reach kinks. The third term, $\tanh(\ell/2x_s)$, arises because adjacent steps compete for available adsorbed molecules.

Exercise 2.14: Case of an isolated step
Find the adsorbed concentration distribution around a single isolated step, and find the step's velocity. [Hint: Either solve the differential equation *da capo* or take the limit of Eqs. (2.31-32) for large ℓ.] ∎

Formula (2.31b), as derived, makes no use of spirals. It is valid for any array of parallel equidistant straight steps, no matter how they are produced. There is yet another way of expressing growth rates if we invoque the spiral mechanism. It involves only the two macroscopic observables: supersaturation and equilibrium bulk concentration. Recall simply that the spiral growth mechanism implied the inverse relation (2.26b). Accordingly, we write

$$\ell/2x_s = \sigma_1/\sigma_r, \tag{2.32a}$$

where σ_1 is a constant. Then, using the equilibrium relation (2.29a) and the definition (2.30b), we get the normal growth rate in the form

$$V(\sigma_r) = (\Omega\sigma_1 KC_e)(\sigma_r/\sigma_1)^2 \tanh(\sigma_1/\sigma_r). \tag{2.32b}$$

This relation, first derived by BCF, is called the "linear-parabolic" law of crystal growth because $V \cong \sigma_r^2$ for small supersaturations (relative to σ_1), whereas that dependence is linear for large supersaturations. Figure 2.10 displays the behavior of Eq. (2.32b), and it is again clear, as in the case of silicon oxidation, that the transition from pure para-

---

† Recall that $z^{-1}\tanh z = 1 + O(z^2)$.

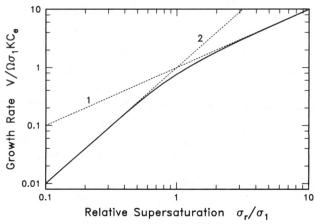

Fig. 2.10: Growth rate *vs.* (relative) supersaturation for the BCF model of crystal growth. Note the limiting linear (1) and quadratic (2) behaviors, shown dashed.

bolic to pure linear behavior occurs over several orders of magnitude in the supersaturation. Thus, the experimental verification of this BCF result is also not easy [36].

There are evidently many other interesting diffusion problems related to the BCF model of crystal growth. In one way or another, these simply relax some of the assumptions (a-e) above. Chernov's excellent review article [37] and textbook [38], and the author's recent paper [39] summarize these extensions. Finally, the exact shape of growth spirals and their dynamics have been investigated by Monte Carlo simulations [40,41].

One parting comment: The BCF theory, in fact, mixes concepts and calculations on a microscopic scale (surface structure) with macroscopic diffusional computations. In that sense it plays the part, in our understanding of crystal growth, of the Boltzmann equation in the kinetic theory of gases. There, the total time derivative (a macroscopic concept) of the single-particle probability density is equal to the "collision integral," which must be computed from scattering theory.

REFERENCES

1. R.P. Smith, "The Diffusivity of Carbon in Iron by the Steady-State Method," *Acta Met.* **1**, 578 (1953).

2. H.S. Carslaw and J.C. Jaeger, *Conduction of Heat in Solids*, 2nd ed., pp. 188-193 (Oxford University Press, London 1959).

3. W.O. Oldham, "The Fabrication of Microelectronic Circuits," *Scientific American* **237**, 111 (1977). [That whole issue is devoted to microelectronics.]

4. B.E. Deal and A.S. Grove, "General Relationship for the Thermal Oxidation of Silicon," *J. Appl. Phys.* **36**, 3770 (1965).

5. E. Rosencher, A. Straboni, S. Rigo, and G. Amsel, "An ^{18}O Study of the Thermal Oxidation of Silicon in Oxygen," *Appl. Phys. Lett.* **34**, 254 (1979).

6. S.I. Raider and R. Flitsch, "X-ray Photoelectron Spectroscopy of SiO_2-Si Interfacial Regions: Ultrathin Oxide Films, *IBM J. Res. & Dev.* **22**, 294 (1978).

7. U.R. Evans, *The Corrosion and Oxidation of Metals* (Edward Arnold Ltd., London 1960).

8. E.A. Irene and Y.J. van der Meulen, "Silicon Oxidation Studies: Analysis of SiO_2 Film Growth Data," *J. Electrochem. Soc.* **123**, 1380 (1976).

9. E.A. Irene and D.W. Dong, "Silicon Oxidation Studies: The Oxidation of Heavily B- and P-Doped Single Crystal Silicon," *J. Electrochem. Soc.* **125**, 1146 (1978).

10. K.J. Laidler, *Chemical Kinetics*, 2nd ed. (McGraw-Hill, New York 1965).

11. R. Ghez and Y.J. van der Meulen, "Kinetics of Thermal Growth of Ultra-thin Layers of SiO_2 on Silicon. Part II: Theory" *J. Electrochem. Soc.* **119**, 1100 (1972).

12. E.A. Irene, "Evidence for a Parallel Path Oxidation Mechanism at the Si-SiO_2 Interface," *Appl. Phys. Lett.* **40**, 74 (1982).

13. E. Bassous, H.N. Yu, and V. Maniscalco, "Topology of Silicon Structures with Recessed SiO_2," *J. Electrochem. Soc.* **123**, 1729 (1976).

14. R.B. Marcus and T.T. Sheng, "The Oxidation of Shaped Silicon Surfaces," *J. Electrochem. Soc.* **129**, 1278 (1982).

15. L.O. Wilson, "Numerical Simulation of Gate Oxide Thinning in MOS Devices," *J. Electrochem. Soc.* **129**, 831 (1982).

16. Proceedings of "Workshop on Oxidation Processes," *Phil. Mag.* **55**, February and June issues (1987).

17. J.W. Gibbs, "On the Equilibrium of Heterogeneous Substances," *Trans. Conn. Acad.* (1875-1878). [Reprinted in *Scientific Papers*, Vol. 1 (Dover, New York 1961).]

18. R. Defay and I. Prigogine, *Tension Superficielle et Adsorption* (Editions Desoer, Liège 1951). [Engl. transl. by D.H. Everett, John Wiley, New York 1966.]

19. R. Ghez, "A Theoretical Crystal Grower's View of Phase Equilibria," in *Int'l School of Crystallography, Erice, 1987* (NATO A.S.I.), to be published (1988).

20. A.J. McConnell, *Applications of Tensor Analysis* (reprinted by Dover, New York 1957).

21. E.A. Guggenheim, *Mixtures* (Oxford University Press, London 1952).

22. A.S. Jordan and M.E. Weiner, "The Effect of the Heat Capacity of the Liquid Phase on the Heat of Fusion Liquidus Equation of Compound Semiconductors," *J. Phys. Chem. Sol.* **36**, 1335 (1975).

23. J.D. Honeycutt and H.C. Anderson, "Small System Size Artifacts in the Molecular Dynamics Simulation of Homogeneous Crystal Nucleation in Supercooled Atomic Liquids," *J. Phys. Chem.* **90**, 1585 (1986).

24. L.E. Scriven, "On the Dynamics of Phase Growth," *Chem. Engng. Sci.* **10**, 1 (1959); **27**, 1753 (1972).

25. F.S. Ham, "Shape-Preserving Solutions of the Time-Dependent diffusion equation," *Quart. Appl. Math.* **17**, 137 (1959).

26. R.F. Sekerka, "Morphologocal Stability," in *Crystal Growth: an introduction*, pp. 403-443, P. Hartman, Ed. (North-Holland, Amsterdam 1973).

27. J.S. Langer, "Instabilities and Pattern Formation in Crystal Growth," *Rev. Mod. Phys.* **52**, 1 (1980).

28. M. Kahlweit, "Precipitation and Aging," in *Physical Chemistry, an Advanced Treatise*, Vol. 10, pp. 719-759, H. Eyring, D. Henderson, and W. Jost, Eds. (Academic Press, New York 1970).

29. I.M. Lifshitz and V.V. Slyozov, "The Kinetics of Precipitation from Supersaturated Solid Solutions," *J. Phys. Chem. Sol.* **19**, 35 (1961).

30. P.W. Voorhees, "The Theory of Ostwald Ripening," *J. Stat. Phys.* **38**, 231 (1985).

31. B. Lewis and G.J. Rees, "Adatom Migration, Capture, and Decay among Competing Nuclei on a Substrate," *Phil. Mag.* **29**, 1253 (1974).

32. J.A. Venables and G.L. Price, "Nucleation of Thin Films," in *Epitaxial Growth*, Part B, pp. 381-436, J.W. Matthews, Ed. (Academic Press, New York 1975).

33. W.K. Burton, N. Cabrera, and F.C. Frank, "The Growth of Crystals and the Equilibrium Structure of their Surfaces," *Phil. Trans. Roy. Soc.* (London) **A243**, 299 (1951).

34. J. Frenkel, "On the Surface Motion of Particles in Crystals and the Natural Roughness of Crystalline Faces," *J. Phys. USSR* **9**, 392 (1945).

35. M. Volmer, *Kinetik der Phasenbildung* (Th. Steinkopff Verlag, Dresden 1939).

36. P. Bennema and G.H. Gilmer, "Kinetics of Crystal Growth," in *Crystal Growth: an introduction*, pp. 263-327, P. Hartman, Ed. (North-Holland, Amsterdam 1973).

37. A.A. Chernov, "The Spiral Growth of Crystals," *Usp. Fiz. Nauk.* **73**, 277 (1961). [Engl. transl. in *Sov. Phys. Usp.* **1**, 116 (1961).]

38. A.A. Chernov, *Modern Crystallography III: Crystal Growth* (Springer-Verlag, Berlin 1984).

39. R. Ghez and S.S. Iyer, "The Kinetics of Fast Steps on Crystal Surfaces and its Application to the Molecular Beam Epitaxy of Silicon," *IBM J. Res. & Dev.*, to be published (1988).

40. G.H. Gilmer, "Growth on Imperfect Crystal Faces," *J. Cryst. Growth* **35**, 15 (1976).

41. G.H. Gilmer, "Computer Models of Crystal Growth," *Science* **208**, 355 (1980).

Chapter 3

DIFFUSION UNDER EXTERNAL FORCES

This short chapter is devoted to the relation between diffusion and mobility. Indeed, until now, the process of mass redistribution was attributed to *mixing* on a microscale. We now ask what effects force fields can have on this redistribution. These affect particle flows through a drift term, in addition to the usual diffusion term.

3.1 ONE-DIMENSIONAL ANISOTROPIC RANDOM WALK

A simple approach consists in relaxing one of the basic assumptions implicit in the random walk model of Chap. 1. There, it was assumed that the jump frequencies were *isotropic*, which means that transition probabilities to the right and to the left had the same value $\frac{1}{2}\Gamma$. We now distinguish between jumps to the right, with jump frequency Γ^+, from jumps to the left, with frequency Γ^-. These frequencies need not be equal; for the moment, we do not ascribe any reason for this asymmetry. As in Sect. 1.1, it follows that the net number of particles that jump from site i to site $i+1$ in unit time is

$$J_{i+1/2} = \Gamma^+ N_i - \Gamma^- N_{i+1}, \tag{3.1a}$$

and the net flux from $i-1$ to i is

$$J_{i-1/2} = \Gamma^+ N_{i-1} - \Gamma^- N_i. \tag{3.1b}$$

These fluxes, now, do not vanish when $N_i = N_{i+1}$, because of the anisotropy in jump frequencies. A constant distribution over sites is therefore no longer a necessary condition for equilibrium. Again, the time rate of change of the particle distribution is evaluated by considering all possible transitions to and from the ith site:

$$\dot{N}_i = \Gamma^- N_{i+1} + \Gamma^+ N_{i-1} - (\Gamma^+ + \Gamma^-)N_i. \tag{3.2}$$

Hence, using Eqs. (3.1), we obtain the same expression (1.3)

$$\dot{N}_i = -(J_{i+1/2} - J_{i-1/2}) \tag{3.3}$$

in terms of fluxes. All the remarks (a-d) of Sect. 1.1 still hold, because we have done nothing except lift the assumption of isotropy of jump frequencies. In particular, Eq. (3.3) still has the form of a conservation law.

We pass to continuum expressions through the same interpolating scheme (1.5-6) as for the isotropic model. Expanding Eqs. (3.2) around the site x_i now yields

$$\frac{\partial \hat{N}}{\partial t} = \frac{1}{2}\Gamma a^2 \frac{\partial^2 \hat{N}}{\partial x^2} - \Delta a \frac{\partial \hat{N}}{\partial x} + O(a^3), \tag{3.4}$$

where, for convenience, we have introduced the sum and difference

$$\Gamma = \Gamma^+ + \Gamma^- \quad \text{and} \quad \Delta = \Gamma^+ - \Gamma^- \tag{3.5}$$

of jump frequencies.

It is again delicate to analyze fluxes. Expanding Eqs. (3.1), we find that

$$J_{i\pm1/2} = \Delta \hat{N} - \frac{1}{2}(\Gamma \mp \Delta)a\frac{\partial \hat{N}}{\partial x}\Big|_{x_i} + O(a^2), \tag{3.6}$$

and, to first order in the jump distance a, these fluxes do *not* converge to the same expression. Since there seems nothing better to do, *define* a continuous flux function

$$\hat{J}(x,t) = \Delta \hat{N} - \frac{1}{2}\Gamma a\frac{\partial \hat{N}}{\partial x} \tag{3.7}$$

as the arithmetic average of the equations (3.6). Note that Eqs. (3.4) and (3.7) both reduce to their homologues (1.7) and (1.9) when the jump frequencies are equal, *i.e.*, when $\Delta = 0$. Then, as in Exercise 1.1, computing the difference $J_{i\pm1/2} - \hat{J}(x_\theta)$ at intermediate points, we find that this difference is $O(a^2)$ when the discrete fluxes are defined at the midpoints $x_{i\pm1/2} = (i\pm\frac{1}{2})a$. Consequently, the conservation law (3.3) has the continuum analog

$$\frac{\partial \hat{N}}{\partial t} = -a\frac{\partial \hat{J}}{\partial x} + O(a^3), \tag{3.8}$$

which, together with Eq. (3.7), is compatible with Eq. (3.4). These equations should be compared with their homologues (1.7-10). Note that the truncation error has increased by one unit because of the jump frequencies' asymmetry.

The required continuum equations emerge if, again, $C = \hat{N}/a$ is the average concentration per cell and if $D = \frac{1}{2}\Gamma a^2$ is the diffusivity. In addition, we define the *drift velocity* $v = \Delta a$, and Eqs. (3.7), (3.8), and (3.4) then become

$$J(x,t) = -D\frac{\partial C}{\partial x} + vC, \tag{3.9}$$

$$\frac{\partial C}{\partial t} = -\frac{\partial J}{\partial x}, \tag{3.10}$$

$$\frac{\partial C}{\partial t} = D\frac{\partial^2 C}{\partial x^2} - v\frac{\partial C}{\partial x}. \tag{3.11}$$

These equations, written in 1-d and for the case of constant coefficients, can be readily generalized. For example,

$$\mathbf{J} = -D\nabla C + \mathbf{v}C \tag{3.12}$$

is the flux vector in an isotropic body of any dimension.

We end this section by inquiring into the physical significance of the drift velocity. Since, by definition, v is proportional to the difference in jump frequencies, it is thus a measure of the *bias* in the system, a bias that subsists even if the particles are evenly distributed among sites. In a multicomponent fluid, one shows [1] that equations similar to Eqs. (3.9-12) describe its dynamics, as was implicit [cf. Eqs. (1.45,48)] in the development of Sect. 1.8. A simple hopping model is then not applicable, but the drift velocity is interpreted simply as the fluid's mean velocity field. In general, the first term of

Eq. (3.12) is called the diffusional flux, while the second, called the convective flux, measures mass transport due to an overall motion. Visualization follows if we think of a dye injected into a running stream.

In a solid there is, in general, no convection because crystal lattices do not move significantly.† Nonetheless, a potential force field $V(\mathbf{x},t)$ can act on particles with a force $\mathbf{F} = -\nabla V$. These fields can be internal (*e.g.*, stresses due to dislocation cores) or caused externally by applied fields (*e.g.*, electric fields that act on charged particles, as in electromigration). Particles that move in a dense medium suffer collisions which tend to randomize their instantaneous velocities. The drift velocity $\mathbf{v}$ is a statistical average over these instantaneous velocities, biased by the force field. One can show [2] that there is often a linear relation

$$\mathbf{v} = B\mathbf{F} = -B\nabla V \tag{3.13}$$

between the drift velocity and the force field, where the coefficient B (necessarily positive) is called the *mobility*. Its dimensions are those of velocity per unit force, or $[B] = \mathrm{s/g}$. We note that Eq. (3.13) is fundamentally different from Newton's law of motion, which relates force to acceleration. Here, because of collisions, the force is proportional to the velocity, as it would be for an overdamped oscillator.

3.2 DIFFUSIVITY AND MOBILITY COEFFICIENTS

Both coefficients D and B characterize the flow of particles. It is therefore not inconceivable that they be related. We shall develop two arguments. The first, due to Mott [3], is based on equilibrium considerations (and is therefore misleading), while the second is dynamical (and is therefore somewhat more believable).

Mott's argument is very simple. The flux (3.12) must vanish identically at equilibrium. With Eq. (3.13), we get

$$-D\nabla C - BC\nabla V = 0. \tag{3.14a}$$

---

† This is not quite true when boundaries move, thermal expansion is taken seriously, or vacancy fluxes are significant (Kirkendall effect).

Multiplying scalarly by an arbitrary displacement $d\mathbf{x}$ gives the differential equation† $D\,dC + BC\,dV = 0$, which has the solution

$$C \propto e^{-BV/D} \qquad (3.14b)$$

if the diffusivity and mobility are constants. But, at equilibrium in a potential field V, classical particles obey a Boltzmann distribution $C \propto \exp(-V/kT)$. Comparing exponents, we get the desired relation

$$D = BkT \qquad (3.15)$$

first suggested by Einstein [4]. Note that the equilibrium concentration (3.14b) cannot be constant if the potential varies with position. The fundamental reason is that only the chemical potential (including fields) must be constant at equilibrium, and not the concentration, which, after all, is but a secondary variable from a basic thermodynamic point of view [5].

Einstein's relation (3.15) was derived, arguably, under equilibrium conditions. Does it hold in general? The next calculation shows it to be true "near" equilibrium. First, we must ask why particles can reside at lattice sites x_i, and why they can hop, occasionally, from site to site. Recall that an isolated particle in a local potential well can be in equilibrium: Any displacement from the potential's minimum position x_i causes a force that restores the particle to its original position. If, however, the potential has finite height, then a sufficiently energetic particle can overcome this barrier, and this particle moves away permanently from its initial position x_i. The site x_i is then metastable because it does not correspond to an absolute minimum of the particle's potential energy.

For an ensemble of particles, such as we are considering (N_i is really a statistical average of the number of particles at the site x_i), the role of potential energy is played by the Gibbs free energy G, and the Gibbs free energy per particle *is* the chemical potential. Referring to Fig. 3.1, we see, for example, that the potential energy is lower at site (2) than at site (1). The transition $1 \rightarrow 2$, if successful, causes a negative change in energy, i.e., $\Delta G = G_2 - G_1 < 0$. In the language of thermodynamics, state (2) is more *stable* than state (1), and ΔG is the free-energy change in the "reaction" $1 \rightarrow 2$. The *rate* of this reaction, however,

---

† Recall from vector analysis that $df = d\mathbf{x} \cdot \nabla f$ for any differentiable function $f(\mathbf{x})$.

Fig. 3.1: Potential-energy diagram showing why the activation barrier ΔG^* differs from the energy change ΔG during reactions.

depends on the potential barrier between these states $\Delta G^* > 0$, which is generally not calculable from thermodynamics alone. Particles at position (more generally, "state") x^* are said to be in a *activated state*, and the barrier ΔG^* is called the *activation energy* [6]. Energy must be imparted, often in the form of heat, to an ensemble of particles so that they can overcome this barrier. In statistical mechanics, one shows [6] that the probability P that a representative particle has an energy at least equal to the activation energy is†§

$$P = e^{-\Delta G^*/kT}. \tag{3.16}$$

Equation (3.16) simply expresses this probability in terms of the ratio of the energy barrier to the thermal energy of particles.

It is now very simple to derive Einstein's relation (3.15). In a crystal lattice, particles reside at sites x_i because they sit in periodic wells, as

---

† In this chapter all energies, including the activation energy, are expressed per particle. Expressions per mole simply require the replacement of Boltzmann's constant k by the gas constant R.

§ Note that all these considerations apply to *any* reaction $1 \rightarrow 2$, such as phase transformations or chemical reactions, for which particles do not necessarily experience displacements. The coordinate x is then called the *reaction coordinate* or the *degree of advancement*.

Fig 3.2: (a) Periodic potential due to a lattice. (b) The bias introduced by the action of an external force **F**.

shown in Fig. 3.2a. These particles can hop from well to well if they have sufficient thermal energy. If v is a characteristic vibration frequency in a given well (approximately the Debye frequency), then

$$\Gamma = 2vP = 2v\,e^{-\Delta G^*/kT} \tag{3.17a}$$

is the rate of *successful* attempts in either direction. If, now, forces **F** act on the particles (directed towards $x > 0$ in Fig. 3.2b), then the potential due to the lattice is decreased by an amount $V = -\int \mathbf{F}\,dx$. For small forces, the maxima are still located at intermediate positions $x_{i\pm1/2}$, but the barriers to the left and right are now different. An estimate similar to (3.17a) gives the jump frequencies

$$\Gamma^\pm = v\,e^{-(\Delta G^* \mp Fa/2)/kT}, \tag{3.17b}$$

where it is assumed that the force's magnitude does not change significantly over a lattice distance. Inserting these expressions into the definitions of diffusivity and drift velocity, we get

$$D = \frac{1}{2}a^2(\Gamma^+ + \Gamma^-) = a^2 \nu e^{-\Delta G^*/kT} \cosh(Fa/2kT), \qquad (3.18a)$$

$$v = a(\Gamma^+ - \Gamma^-) = 2a\nu e^{-\Delta G^*/kT} \sinh(Fa/2kT), \qquad (3.18b)$$

and, from the definition (3.13) of the mobility, the ratio

$$B/D = (2/Fa)\tanh(Fa/2kT). \qquad (3.18c)$$

This ratio tends towards $(kT)^{-1}$ when the potential drop due to the force field is small compared to the thermal energy,† and we recover Einstein's relation (3.15) in this limit.

Finally, the transcription of these relations to charged particles is instructive. Let q be the charge (including the sign) of the particles in question. The force on these particles due to an electric field is $\mathbf{F} = q\mathbf{E}$, and therefore the drift velocity is

$$\mathbf{v} = Bq\mathbf{E} \qquad (3.19a)$$

One calls

$$\mu = B|q| \qquad (3.19b)$$

the *mobility coefficient* for electronic or ionic systems. It is defined here as a positive quantity so as to recover the semiconductor fluxes (1.40c,d), and the symbol μ should not be confused with the same symbol used for chemical potentials. Note that its dimensions are $[\mu] = s\,Cb/g = cm^2/V\,s$. With these definitions, Eqs. (3.13) and (3.15) read‡

$$\mathbf{v} = \text{sgn}(q)\mu\mathbf{E} \quad \text{and} \quad D = \mu kT/|q|. \qquad (3.19c,d)$$

In particular, the flux (3.12) for charged particles is

$$\mathbf{J}(x,t) = -D\nabla C + \text{sgn}(q)\mu C\mathbf{E}. \qquad (3.20)$$

---

† Recall that $\tanh z = z + O(z^3)$.

‡ The *signum* function $\text{sgn}(z)$ has values ± 1 according to the sign of its argument z.

Exercise 3.1: *Concentration distribution in a slab*

Consider a dielectric, such as the SiO_2 layer of Fig. 2.2. Find the steady-state distribution of charged particles in that region, assuming that only one charged species is mobile, that the electric field is constant throughout the dielectric, and that the interface concentrations are known. Discuss for both positive and negative charges. ∎

The points we have broached are presently the object of active investigations [5,7,8] into the nature of fluctuations around, and the stability of, the steady state. Moreover, Eqs. (3.18) show that both D and B are functions of location if the external force F is nonconstant. They are said to be "state dependent." Conversely, a very recent paper [9] shows how a given state-dependent diffusivity is equivalent to a bias in the potential energy and thus to a nonvanishing flux.

3.3 AN INTRODUCTION TO DOUBLE LAYERS

The last exercise was designed to illustrate that charged particles in 1-d steady state are generally *not* distributed linearly. This is due to two competing influences on the flux (3.20): The diffusion term promotes mixing and thus leads to increased randomness or entropy, while the second term represents the system's attempt to lower its energy through the electric field. This field, in turn, depends on the *actual* distribution of charges, as well as on externally applied bias voltages. Problems of this nature are thus nonlinear, and their solution depends heavily on the charges' behavior near surfaces. In general, we can expect charge imbalances close to phase boundaries, *i.e.*, local charge neutrality breaks down. This is called a *double layer* or *space charge* because a net distributed charge must be compensated by a surface charge of opposite sign, thus giving rise to an electric dipole. Double layers are very common: Electrolytes in batteries, electrons and holes in semiconductors, and ions in dielectrics all share their features.

Here, we examine a particularly simple case that displays, nevertheless, all the generic difficulties of double layers. Consider, as shown in Fig. 3.3, a semi-infinite, one-dimensional dielectric, of dielectric constant ε, that is terminated by a metal electrode. Assume, for example, that only one singly charged ionic species is mobile and that its charge e

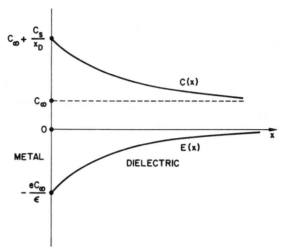

Fig. 3.3: Concentration and electric-field distributions in a dielectric. Note the charge separation within a Debye length of the interface.

is positive, say. The ions' concentration distribution C in the dielectric is compensated by a fixed negative background at a concentration level C_∞. The ions are positive because they have discharged electrons at the electrode. Therefore, there is an induced negative charge on the metal whose surface concentration is C_s. We then wish to compute the steady-state distribution of these positive ions and their electric field or potential. In accordance with Sects. 2.1-2, the assumption of steady state in 1-d implies that the particle flux (3.20) is constant with respect to location x. If, further, we assume that the electrode is "blocking," meaning that J vanishes there, then it follows that

$$-D\frac{dC}{dx} + \mu CE = 0 \qquad \forall x > 0, \tag{3.21a}$$

with $\mu/D = e/kT$ according to Einstein's relation (3.19d) . The distributed charge density, however, is a source of electric field, given by Gauss's law

$$\varepsilon\frac{dE}{dx} = e(C - C_\infty), \tag{3.21b}$$

and the negative surface charge also gives rise to the field at the metal-dielectric phase boundary

$$\varepsilon E(0+) = -eC_s. \qquad (3.22a)$$

Finally, electrical neutrality must hold far from the electrode

$$C(\infty) = C_\infty. \qquad (3.22b)$$

We must now solve the nonlinear system of two first-order differential equations (3.21) for the unknown functions $C(x)$ and $E(x)$, under the boundary conditions (3.22). Before doing this, several consequences can be drawn from these equations. First, the bulk neutrality condition (3.22b), together with Gauss's law (3.21b), implies that

$$E'(\infty) = 0, \qquad (3.23a)$$

where primes denote differentiation with respect to distance x. Next, we can formally integrate Eq. (3.21b) with the boundary condition (3.22a) to give

$$\varepsilon E(x) = -eC_s + e\int_0^x [C(\xi) - C_\infty]d\xi.$$

This merely states that the field at any location x is proportional to the sum of charges contained in the region to the left of x. For large x this region contains *all* charges, and, since global neutrality must hold, it follows that

$$E(\infty) = 0. \qquad (3.23b)$$

Therefore, both the electric field and its gradient must vanish at infinity. The solution of Eqs. (3.21) follows by eliminating C in favor of the field

$$C = C_\infty + (\varepsilon/e)E', \qquad (3.24a)$$

which, inserted into Eq. (3.21a), yields the second-order differential equation

$$E'' - \frac{e}{kT}E\left(E' + \frac{eC_\infty}{\varepsilon}\right) = 0. \qquad (3.24b)$$

Since this equation does not explicitly depend on x, the introduction of a new dependent variable $u = E'$, considered a function of E, transforms it into the first-order equation

$$u\frac{du}{dE} - \frac{e}{kT}(u + eC_\infty/\varepsilon)E = 0 \qquad (3.25a)\dagger$$

whose variables separate. With the subsidiary conditions (3.23), we find that

$$\frac{\varepsilon}{eC_\infty}E' - \ln\left(1 + \frac{\varepsilon}{eC_\infty}E'\right) = \frac{\varepsilon}{2kTC_\infty}E^2. \qquad (3.25b)$$

And this is as far as we can go analytically without further approximations because this equation gives E' implicitly. If, following Debye and Hückel [10], we assume that the field is "small" in the sense that

$$|(\varepsilon/eC_\infty)E'| \ll 1, \qquad (3.26)$$

then the logarithm can be expanded to second order in that quantity‡ to yield the equation

$$x_D E' = \pm E, \qquad (3.27a)$$

where the characteristic distance

$$x_D = \sqrt{\varepsilon kT/e^2 C_\infty} \qquad (3.27b)$$

is called the Debye length. It measures the extent of the double layer as the following shows. Equation (3.27a) is readily solved under the boundary conditions (3.22a) and (3.23a) to give

$$E(x) = -\frac{eC_s}{\varepsilon} e^{-x/x_D}, \qquad (3.28a)$$

and the concentration distribution follows from Eq. (3.24a) by differentiation

$$C(x) = C_\infty + (C_s/x_D) e^{-x/x_D}. \qquad (3.28b)$$

Thus, both C and E decay exponentially into the dielectric according to the Debye length.

---

† Use the identity $E'' = du/dx = (du/dE)(dE/dx) = u(du/dE)$.
‡ Recall that $\ln(1+z) = z - z^2/2 + O(z^3)$.

Exercise 3.2: *When is the field small?*
First, carefully justify the steps that lead from Eqs. (3.24) to (3.28). Then, express the condition (3.26) in terms of material parameters, and interpret. Find the dipole moment. ∎

A few remarks are in order. First, it is clear that we could have obtained the same results by eliminating the field E rather than C. (How would you do this?) Next, a general solution of the first integral (3.25b) requires numerical methods. Further, the above calculation did not require the introduction of the electrostatic potential $\phi = -\int E \, dx$. Multidimensional problems, on the other hand, are more easily solved in terms of ϕ. We notice, however, as with Mott's argument (3.14), that Eq. (3.21a) can be integrated in terms of the potential to give Boltzmann's distribution $C \propto e^{-e\phi/kT}$. Finally, several mobile charges can be considered, such as electrons and holes, donors and acceptors in semiconductors [11]. Likewise, oxidation kinetics in metals [12] often depend on the motion of both ions and electrons. In all these cases, the surface charge plays a crucial part, and the identification and control of *surface states* due to dangling bonds or to adsorbed ions is of the utmost importance [13-15].

Exercise 3.3: *The case of uncompensated bulk charge*
When the negative background C_∞ is negligible, the distributed charge is entirely compensated by the surface charge. Solve that corresponding problem (3.21-22) exactly, and define an appropriate Debye length. ∎

REFERENCES

1. R.B. Bird, W.E. Stewart, and E.N. Lightfoot, *Transport Phenomena*, Chap. 18 (John Wiley, New York 1960).
2. E.M. Lifshitz and L.P. Pitaevskii, *Physical Kinetics*, (Vol. 10 of *Course on Theoretical Physics*) pp. 40-42 (Pergamon Press, Oxford 1981).
3. N.F. Mott and R.W. Gurney, *Electronic Processes in Ionic Crystals*, 2nd ed., p. 63 (reprinted by Dover, New York 1964).
4. A. Einstein, papers on brownian motion, collected in *Investigations on the Theory of the Brownian Movement* (reprinted by Dover, New York 1956).
5. R. Landauer, "Stability and Relative Stability in Nonlinear Driven Systems," *Helv. Phys. Acta* **56**, 847 (1983).

6. A.S. Glasstone, K.J. Laidler, and H. Eyring, *The Theory of Rate Processes* (McGraw-Hill, New York 1941).

7. R. Landauer, "The ballast resistor," *Phys. Rev.* **A15**, 2117 (1977).

8. R. Landauer, "Stability in the Dissipative Steady State," *Physics Today* **31**, 23 (Nov. 1978).

9. M. Büttiker, "Transport as a Consequence of State-Dependent Diffusion," *Z. Physik* **B68**, 161 (1987).

10. J. Frenkel, *Kinetic Theory of Liquids*, pp. 36-40 (reprinted by Dover, New York 1955).

11. E. Spenke, *Electronic Semiconductors* (McGraw-Hill, New York 1958).

12. A.T. Fromhold, *Theory of Metal Oxidation*, Vols. I and II (North-Holland, Amsterdam 1976, 1980).

13. J. Bardeen, "Surface States and Rectification at a Metal Semi-conductor Contact," *Phys. Rev.* **71**, 717 (1947).

14. V. Heine, "Theory of Surface States," *Phys. Rev.* **138**, A1689 (1965).

15. A. Many, Y. Goldstein, and N.B. Grover, *Semiconductor Surfaces* (North-Holland, Amsterdam 1965).

Chapter 4

SIMPLE TIME-DEPENDENT EXAMPLES

Our ability to solve certain differential equations increases with our lexicon of available functions. Exponentials, for example, are basic solutions of ordinary differential equations with constant coefficients. Likewise, to solve time-dependent problems in diffusion theory requires the introduction of certain "special functions" that are closely related to gaussians [1,2]. In this chapter we follow a constructive path that issues from our earlier experience with random walks, through new physical examples, and finally to the family of error functions. We then show, in the next chapter, that these error functions are, in fact, part of a grander scheme — similarity — that also includes the classical method of separation of variables. This chapter ends with a consideration of problems in several space dimensions that can be expressed in terms of products of 1-d problems.

4.1 THE GAUSSIAN AND ONE OF ITS RELATIVES

Let us return to the isotropic random walk and to its numerical solution presented in Sects. 1.4-5. It was suggested there that the reader plot and fit the output of the program $EXPLICIT$, much as one would with experimental data. Figure 4.1 shows the resulr for the benefit of those who did not take the suggestion. Here, $\ln C_i^n$ is plotted against the square of the space index i. Except around the origin, we obtain essentially straight lines $\ln C_i^n \cong \alpha - \beta i^2$ with negative slopes. The absolute value β of these slopes appears to decrease with increasing time index n, as do the intercepts α. In fact, further data fitting suggests that β is inversely proportional to n, but that the intercepts are slowly varying with time. More precisely, one finds that $\beta = (4\lambda n)^{-1}$, and we need not even find $\alpha(n)$. Exponentiation of the above logarithmic relation

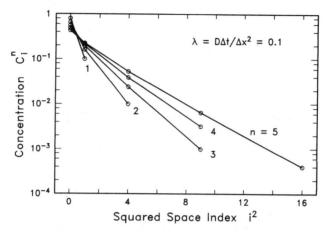

Fig. 4.1: Semilog plot of output from the explicit finite-difference scheme of Sect. 1.5.

yields

$$C_i^n \propto e^{-i^2/4n\lambda}, \tag{4.1a}$$

and, remembering the definitions (1.24) and (1.21) of the modulus and grid points, we obtain the continuum form

$$C(x,t) = A\,e^{-x^2/4Dt}. \tag{4.1b}$$

The coefficient of proportionality A is easily obtained, since we know that there was unit mass to begin with, and that mass is conserved by diffusion in an infinite medium. Therefore we must have

$$\int_{-\infty}^{\infty} C(x,t)\,dx = 1 \tag{4.1c}$$

for all times. Inserting the expression (4.1b) into Eq. (4.1c) and recalling [3, p. 371] a standard result

$$\int_{-\infty}^{\infty} e^{-\xi^2}d\xi = \sqrt{\pi}, \tag{4.2}$$

we get the normalization constant $A = (4\pi Dt)^{-1/2}$ required in Eq. (4.1b). In sum, the continuum solution of the diffusion equation for a point source in an infinite medium is the gaussian

$$G(x,t) = \frac{1}{2\sqrt{\pi Dt}} e^{-x^2/4Dt}, \tag{4.3}$$

and Fig. 1.3 of Chap. 1 is its graph.

The resemblance between this solution (4.3) and the normal distribution of probability theory [4] is not fortuitous. To grasp this connection, recall that the explicit finite-difference formula (1.23a)

$$C_i^{n+1} = (1 - 2\lambda)C_i^n + \lambda(C_{i+1}^n + C_{i-1}^n), \tag{4.4}$$

repeated here for convenience, samples the actual concentration distribution at grid points (x_i, t_n). But that formula can also be interpreted in terms of a random walk if the sampled concentration C_i^n represents the occupation probability of site i at the time step n, and if λ represents the *a priori* (conditional) probability of a jump (per unit time) either to the right or to the left. Indeed, there are three mutually exclusive events that lead to the occupation probability on the left-hand side of Eq. (4.4): a jump to the right from state $(i-1,n)$, a jump to the left from state $(i+1,n)$, and no jump at all from state (i,n). These transitions are represented by arrows in Fig. 1.4, and they correspond to the computational molecule of the finite-difference scheme. Thus, after n throws of a "three-sided coin," the probable positions of a particle that was initially at the origin, say, will be given by a trinomial distribution characterized by the three *a priori* probabilities λ, λ, and $1-2\lambda$.† As the number of trials increases, one shows that this distribution converges,‡ like the binomial distribution, to the gaussian (4.3). In fact, the stability criterion (1.26) simply means that the *a priori* probabilities cannot be negative.

The gaussian (4.3) is a major building block for the construction of time-dependent solutions. It sometimes goes under the name of *fundamental solution* or *Green's function* (in an infinite domain). From its heuristic derivation, above, we have the following properties which the reader can easily verify:

---

† In fact, the special case $\lambda = 1/2$ reduces Eq. (4.4) to a recursion that is very close to the one that defines Pascal's triangle. The *a priori* probability of no jump is then zero.

‡ This is, in fact, a special case of the "central limit theorem" which pervades probability theory.

a. It is a solution of the diffusion equation (1.14).

b. It vanishes at infinity ($x \rightarrow \pm\infty$) for all positive values of time t.

c. Its integral over all space is precisely 1.

d. It tends to zero as t tends to zero,[†] except at the origin $x = 0$, where it is singular. This, together with the last property, is characteristic of the delta function. Thus we have the representation $\lim_{t \to 0+} G(x,t) = \delta(x)$, which is nothing but the initial condition of the problem solved numerically in Sect. 1.5.[‡]

These properties are valid for higher dimensions, as shown in the next

Exercise 4.1: *Fundamental solution in d dimensions*
Show that

$$(4\pi Dt)^{-d/2} e^{-r^2/4Dt}$$

is a solution of the d-dimensional diffusion equation (1.33) if r is the length of the position vector. Verify also Properties (b-d). [Hint: It is easiest to verify Property (a) in rotationally symmetric coordinates [see Eq. (1.36)], but the other proofs are more transparent in cartesian coordinates.] ∎

As a first example of the gaussian's use, we now give the continuum analog of the result (1.29b): An arbitrary initial distribution $\varphi(x)$ can be viewed as the superposition of point sources. It will thus evolve with time according to

$$C(x,t) = \int_{-\infty}^{\infty} G(x-x',t)\,\varphi(x')\,dx', \tag{4.5}$$

where G is given by the function (4.3). The reasoning is exactly the same as in the discrete case. It is also easy to verify directly that Eq. (4.5) satisfies the diffusion equation and the boundary conditions at infinity. The verification of the initial condition is a consequence of the delta-function property of G.§

---

† This convergence to zero is faster than that of any power of t; this is called an *essential singularity*.

‡ We will have more to say on delta functions in Sect. 6.2 of Chap. 6.

§ A more formal proof, but by no means a rigorous one, follows from the change of variables $x' = x + 2\sqrt{Dt}\,\xi$ in Eq. (4.5).

The elegance of a formula such as Eq. (4.5) can be appreciated if we recognize that it produces, explicitly, the concentration distribution at all locations and all times through a simple integration over an arbitrarily given initial configuration. The boundary conditions are built in. The initial conditions might even be given numerically, as with experimentally determined data. For example, if diffusion occurs only in a half space $\{x > 0\}$, then we find the surface concentration by setting $x = 0$ in Eq. (4.5), even *before* performing the integral. [What should be the boundary condition if G is the gaussian (4.3)?] For each set of boundary conditions there exists a Green's-function representation such as Eq. (4.5). In Chaps. 6 and 7, we shall see that the Laplace convolution theorem allows us to find these representations, almost automatically.

Exercise 4.2: A gaussian remains a gaussian
An arbitrary initial gaussian $(Q/\sigma\sqrt{\pi})\exp[-(x-m)^2/\sigma^2]$ is entirely characterized by three constants: Q, m, σ. Show that it preserves its shape under diffusion in an infinite medium. Note how the mean, maximum, and variance (or dispersion) change with time. This problem is of some concern to those who use ion implantation; *cf.* [5]. ∎

Consider now an initial distribution of the form

$$\varphi(x) = \begin{cases} -1 & \text{if } x < 0, \\ 1 & \text{if } x > 0. \end{cases} \tag{4.6a}$$

This nonphysical distribution (part of φ is negative) leads us quickly down the path of error functions. Inserting the function (4.6a) into (4.5) and changing variables as in the last footnote, we get the expression

$$C(x,t) = \frac{2}{\sqrt{\pi}} \int_0^{x/2\sqrt{Dt}} e^{-\xi^2} d\xi. \tag{4.6b}$$

The integral on the right-hand side cannot be reduced to elementary functions. It is so common, however, that it is given a name meant to remind us of its origin in probability theory.

Definition: The error function is defined by the integral†

$$\text{erf}(z) = \frac{2}{\sqrt{\pi}} \int_0^z e^{-\xi^2} d\xi. \tag{4.7}$$

With this definition, the solution to our problem (4.6) is simply

$$C(x,t) = \text{erf}(x/2\sqrt{Dt}). \tag{4.6c}$$

The error function has the following properties, which the reader should verify directly from the definition (4.7) to gain familiarity with this important function:

a. It is an odd function, *i.e.*, $\text{erf}(-z) = -\text{erf}(z)$, which implies that $\text{erf}(0) = 0$.

b. Its limit $\lim_{z \to \infty} \text{erf}(z) = 1$, and, with Property (a), its limit as $z \to -\infty$ is -1.

c. The differentiation formula $(d/dz)\text{erf}(z) = (2/\sqrt{\pi})\exp(-z^2)$.

d. A Taylor expansion of the exponential and termwise integration yield the series representation

$$\text{erf}(z) = \frac{2}{\sqrt{\pi}} \sum_{n=0}^{\infty} \frac{(-1)^n z^{2n+1}}{n!(2n+1)}.$$

e. The function (4.6c) satisfies the diffusion equation under the boundary conditions

$$C(0,t) = 0 \quad \text{and} \quad C(\infty,t) = 1 \quad \text{for } t > 0,$$

and the initial condition

$$C(x,0) = 1 \quad \text{for } x > 0.$$

What does all this mean? First, the parity of $\text{erf}(z)$ follows from the gaussian's evenness. Both are represented in Fig. 4.2, and it is evident from the definition that $\text{erf}(0) = 0$. This is also clear from our method of construction, in which an odd function (4.6a) was allowed to relax by diffusion. Next, the crucial Property (e) follows, in fact, from the invariance properties C.1 in Sect. 1.3, because the error

---

† To avoid confusion with x and t or any combination thereof, the variable z in this definition (and others to follow) represents any real variable. In fact, most of the following expressions in z can be extended to the complex domain.

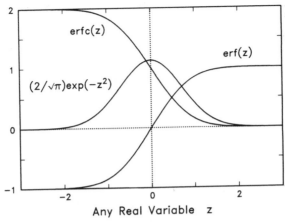

Fig 4.2: Graph of the gaussian and error functions.

function (4.6c) is essentially an integral over space of the gaussian (4.3), which is itself a solution of the diffusion equation. Finally, the integration (4.5) was performed through a point of discontinuity (the origin $x = 0$) of the initial distribution φ. This is reflected in the "time-developed" solution (4.6c), which again has a singularity at the origin of the (x,t)-plane. In that connection, we recall that any function of two variables, such as the concentration distribution $C(x,t)$, is equivalent to a surface. (Remember Fig. 1.3 for the gaussian.) Rather than suffer with perspective drawings or from incomplete information given by isochronal profiles, it is often convenient to represent the behavior of such functions directly on the (x,t)-plane of the independent variables. Figure 4.3 shows how the error function (4.6c) behaves on the axes and at infinity. The figure clearly demonstrates that simple error function solutions require *constant* initial and boundary conditions on each half axis, and the initial conditions must match the far-field conditions. The origin is evidently a point of discontinuity that must be excluded from the initial and boundary conditions. It is precisely this discontinuity that propagates forward in time — an infection, so to speak — to produce the actual concentration distribution.

The error function has many applications. For example, consider the simple problem of *impurity diffusion in a semi-infinite region.* For definiteness, assume that a dopant is initially distributed in a silicon

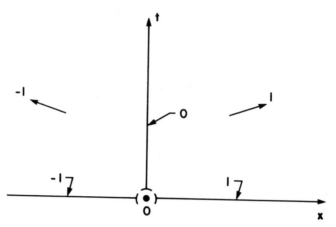

Fig. 4.3: Domain of the independent variables and particular values of the error-function solution of the diffusion equation. Note the singular behavior at the origin.

wafer at a constant concentration level C_∞. At time $t = 0$, this wafer is placed in a reactive atmosphere inside a furnace. Assume, also, that the dopant reacts with this gas in such a way that local equilibrium is maintained at the wafer-gas surface, taken as the origin of coordinates. The dopant's concentration there will then be a constant $C_e < C_\infty$ that depends only on the reactant's temperature and pressure. We want to know at what rate the dopant escapes from the wafer. The solution is immediate if we scale the concentration properly. In fact, the function

$$u(x,t) = \frac{C(x,t) - C_e}{C_\infty - C_e} \tag{4.8a}$$

satisfies all the conditions of Property (e) above. Therefore,

$$C(x,t) = C_e + (C_\infty - C_e)\,\mathrm{erf}(x/2\sqrt{Dt}) \tag{4.8b}$$

is the required solution. Note that we have used Property C.1 of Sect. 1.3 in the sense that an arbitrary constant is a (trivial) solution of the diffusion equation. Next, the rate at which dopant leaves the semi-conductor is simply proportional to the surface flux J_0. Using Property (c), we easily get

$$J_0 = -DC_x \big|_{x=0} = -(C_\infty - C_e)\sqrt{D/\pi t}, \tag{4.8c}$$

which displays a characteristic $t^{-1/2}$ behavior. In addition, the minus sign reminds us that the flux is directed into the gas if the x-axis points into the Si. Finally, the solution (4.8b) is valid for any relative values of C_e and C_∞. That solution, therefore, can describe equally well the "indiffusion" into a semi-infinite body of an impurity from a constant external source. Problems of this type have no constant physical length scale, but we do note that the solution (4.8b) scales distances x with $\sqrt{Dt}$. This quantity, for reasons that will soon become clear, is called the *diffusion distance*.

The novice will perhaps be disturbed by the foregoing procedure. It may appear that we have pulled a solution from a hat. The constructive method, first mentioned in the introduction to this chapter, consists essentially in taking known solutions of the diffusion equation, perhaps rescaled, and trying to match the given initial and boundary conditions. There are two possibilities: Either we are unsuccessful and we must then look for other known solutions, or the conditions can all be met. Thus, by virtue of the uniqueness theorem mentioned at the end of Sect. 1.3 of Chap. 1, we are sure to have found *the* solution to our problem, because there can be none other.

4.2 THE GAUSSIAN AND ANOTHER OF ITS RELATIVES

To motivate the equally important *complementary error function* let us return to *Stokes's first problem*, last mentioned in Sect. 1.7. Together, Eqs. (1.38) show that the velocity distribution $v(x,t)$ must satisfy the diffusion equation

$$\frac{\partial v}{\partial t} = \nu \frac{\partial^2 v}{\partial x^2}. \tag{4.9a}$$

Assume that the flat plate impulsively moves in its plane at a constant velocity V. The fluid is initially at rest, but, because it "sticks" at material boundaries, its tangential velocity there must equal the plate's: $v(0,t) = V$. But the function $1 - v/V$ satisfies all the requirements of Property (e) of the last section. The solution is therefore

$$v(x,t) = V\left[1 - \text{erf}(x/2\sqrt{\nu t})\right]. \tag{4.9b}$$

The function in brackets occurs so frequently that one feels compelled to introduce the

Definition: The complementary error function is defined by

$$\text{erfc}(z) = 1 - \text{erf}(z). \tag{4.10a}$$

From this definition, the standard integral (4.2), and the definition (4.7), we get the integral representation

$$\text{erfc}(z) = \frac{2}{\sqrt{\pi}} \int_z^\infty e^{-\xi^2} d\xi. \tag{4.10b}$$

The solution to Stokes's problem can therefore be expressed as

$$v(x,t) = V\,\text{erfc}(x/2\sqrt{\nu t}). \tag{4.9c}$$

The complementary error function also enjoys properties similar to the ordinary error function. They are easy to verify:

a. It is an odd function around the point $(0,1)$, which implies that $\text{erfc}(0) = 1$.

b. Its limit for large positive z is $\lim_{z \to \infty} \text{erfc}(z) = 0$, and, with Property (a), its limit as $z \to -\infty$ is 2. The symmetry property thus reads $\text{erfc}(-z) = 2 - \text{erfc}(z)$.

c. The differentiation formula $(d/dz)\text{erfc}(z) = -(2/\sqrt{\pi})\exp(-z^2)$.

d. Its Taylor expansion is simply related to Property (d) of $\text{erf}(z)$ through the definition (4.10a).

e. The function $C(x,t) = \text{erfc}(x/2\sqrt{Dt})$ satisfies the diffusion equation under the boundary conditions

$$C(0,t) = 1 \quad \text{and} \quad C(\infty,t) = 0 \quad \text{for } t > 0,$$

and the initial condition

$$C(x,0) = 0 \quad \text{for } x > 0.$$

These conditions are evidently complementary (and thus the name) to those satisfied by the ordinary error function. This function is also displayed in Fig. 4.2. Note that it looks like a metallurgical junction. We shall soon see that this appearance is not misleading.

Another useful property of the complementary error function is the way in which it approaches zero for large *positive* argument (in practice, $z \gtrsim 2$). This fast decay is interesting analytically, and it is also useful for numerical calculations. This leads us gently to the notion of an *asymptotic expansion*, a notion that generalizes the asymptotes of high-school geometry. To simplify the calculations (it is always easier to work with simple exponentials) make the change of variables $\alpha = \xi^2$ in Eq. (4.10b), and integrate by parts:

$$\mathrm{erfc}(z) = \frac{1}{\sqrt{\pi}} \int_{z^2}^{\infty} \alpha^{-1/2} e^{-\alpha} \, d\alpha$$

$$= \frac{1}{\sqrt{\pi}} \left[z^{-1} e^{-z^2} - \frac{1}{2} \int_{z^2}^{\infty} \alpha^{-3/2} e^{-\alpha} \, d\alpha \right],$$

and the process can be repeated to yield a formal series

$$\sqrt{\pi} \, z e^{z^2} \mathrm{erfc}(z) = 1 - \frac{1}{2z^2} + \frac{1 \times 3}{(2z^2)^2} - \ldots\ldots \tag{4.11a}$$

We seem to have generated a series in inverse powers of the argument z, and each term must thus be smaller than its predecessors as $z \to \infty$. This series, however, is not convergent, as is easily seen with the ratio test. In fact, for fixed z, the successive terms of this series will eventually increase. Nevertheless, a series such as given by Eq. (4.11a) is useful because the partial sums approach the function in question, with ever-increasing accuracy, as z increases. Using a tilde to symbolize an asymptotic relation, Eq. (4.11a) reads

$$\mathrm{erfc}(z) \sim \frac{e^{-z^2}}{z\sqrt{\pi}} \left[1 + \mathrm{O}(z^{-2}) \right] \quad \text{for } z > 0, \tag{4.11b}$$

and in practice this is all we will ever need.

More, much more, on asymptotic expansions can be found in Erdélyi's monograph [6], and Lin and Segal [7] give an excellent introductory account of the theory and its applications. For example, one shows that the error committed in retaining N terms of an asymptotic expansion is less, in absolute value, than the last (the Nth) included term. For our purposes, however, Eqs. (4.11) show that the complementary error function behaves like a gaussian divided by the argument z to the first power, at least for large values of that argument.

One says that it has a gaussian "tail." How large is "large" depends on what we mean by "small" values of the gaussian. For example, for $z = 2$ it follows from Eq. (4.11b) that the complementary error function is of the order of 1% of its value (unity) at the origin. Such estimates allow operational definitions of "diffusion distances," "penetration lengths," or "boundary-layer thicknesses" — all equivalent terminology. For example, because our previous solutions (4.8b) and (4.9c) contain distance and time only in the particular combination $z = x/2\sqrt{Dt}$, it follows that the fields have decayed to 1% of their boundary values at distances of the order of $4\sqrt{Dt}$.

All this is fine and well, but how do we *compute* error functions? Of course, one can call an appropriate subroutine or evaluate the integrals (4.7) and (4.10b) numerically. Table 4.1, however, shows an excellent rational fraction approximation that is easily programmed (an APL function is also listed) and that could always help out on a desert

Table 4.1: Algorithm and listing of an APL function for the evaluation of error functions

For non-negative x, $\mathrm{erfc}(x) = e^{-x^2}\dfrac{P(x)}{Q(x)}$, where $P(x) = \sum\limits_{n=1}^{7} P_n x^n$ and $Q(x) = \sum\limits_{n=1}^{8} Q_n x^n$.

The coefficients of these polynomials are:

$P_1 =$	440.4137358247522000	$Q_1 =$	440.41373582475220
$P_2 =$	625.6865357696830000	$Q_2 =$	1122.64022017704700
$P_3 =$	448.3171659914834000	$Q_3 =$	1274.66726675766200
$P_4 =$	191.8401405879669000	$Q_4 =$	838.81036547840640
$P_5 =$	50.9916976282532000	$Q_5 =$	347.12292881178430
$P_6 =$	7.9239298287415450	$Q_6 =$	90.87271120353703
$P_7 =$	0.5641999455982575	$Q_7 =$	14.04533730546114
		$Q_8 =$	1.00000000000000

For negative x, use $\mathrm{erfc}(x) = 2 - \mathrm{erfc}(|x|)$.

```
      ∇  Z←ERFC X
[1]   A    EVALUATES ERFC(X) TO 15D FOR 0 ≤ |X| ≤ 10
[2]      Z←((ρX),1)ρ|X
[3]      Z←|(2×X<0)-(*-X×X)×(Z⊥P)÷Z⊥Q
      ∇
```

island. The curves in Fig. 4.2, in fact, were evaluated with this function. It gives 15 significant digits in the range $0 \le |x| \le 10$. For negative argument we need only use the symmetry property (b) and, for large positive argument, the asymptotic expansion (4.11b).

Exercise 4.3: *A few integrals*
Gaussians and error functions are widely used to express certain integrals. For real λ, show that:

1. $\dfrac{2}{\sqrt{\pi}} \displaystyle\int_0^\infty e^{-\xi^2} e^{-2\lambda\xi} d\xi = e^{\lambda^2} \mathrm{erfc}(\lambda).$

[Hint: Complete the square.]

2. $\dfrac{2}{\sqrt{\pi}} \displaystyle\int_0^\infty e^{-\xi^2} \cos(2\lambda\xi) d\xi = e^{-\lambda^2}.$

[Hint: Extend the integral over the range $[-\infty,\infty]$ and reduce to the previous problem.]

3. $\dfrac{2}{\sqrt{\pi}} \displaystyle\int_0^\infty e^{-\xi^2} e^{-(\lambda/\xi)^2} d\xi = e^{-2|\lambda|}.$

[Hint: Change variables $\xi \to \eta = \lambda/\xi$ and form a differential equation.] ∎

4.3 TWO APPLICATIONS OF ERROR FUNCTIONS TO ONE-DIMENSIONAL PHASES

We have just learned that error functions are particular solutions of the diffusion equation — particular in the mathematical sense that they satisfy given constant initial and boundary conditions *on the field*, as opposed to conditions on its derivatives. The impurity diffusion problem (4.8) and Stokes's problem (4.9) were both of this type. In this section, we apply these functions to the solution of problems that are just slightly more complicated than those just mentioned.

Our first problem will be to compute the *time-dependent temperature field in a 1-d, infinite, two-phase system*. Specifically, consider two large and different chunks of material, labeled (1) and (2), each initially at constant temperature $T_\infty^{(\alpha)}$, $\alpha = 1,2$. Now bring these materials together along a common plane phase boundary, and let heat flow from the hotter body to the colder.† We ask how the temperature field varies with position and time, and, in particular, what is the temperature T_0 at the interface $x = 0$, say, joining these two phases.

The initial and boundary conditions for this problem are represented in Fig. 4.4, which is similar to the previous Fig. 4.3, and the temperature field must satisfy heat equations (1.37)

$$\frac{\partial T^{(\alpha)}}{\partial t} = \kappa^{(\alpha)} \frac{\partial^2 T^{(\alpha)}}{\partial x^2}, \quad \alpha = 1, 2 \tag{4.12}$$

in phase (1) $\{x < 0\}$ and in phase (2) $\{x > 0\}$. Because the complementary error function vanishes at $+\infty$, we immediately write the general solution in the form

$$T^{(1)}(x,t) = T^{(1)}_{\infty} + A^{(1)}\mathrm{erfc}[-x/2(\kappa^{(1)}t)^{1/2}], \tag{4.13a}$$

$$T^{(2)}(x,t) = T^{(2)}_{\infty} + A^{(2)}\mathrm{erfc}[x/2(\kappa^{(2)}t)^{1/2}], \tag{4.13b}$$

where the A's are constants yet to be determined. Because of Property (e) of the last section, the functions (4.13) certainly satisfy the equations (4.12) and the initial and far-field conditions. Note that

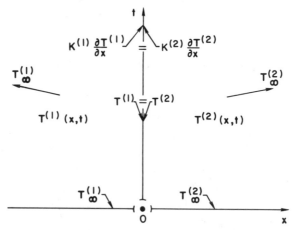

Fig. 4.4: Domains of the independent variables and auxiliary conditions for the two-phase thermal redistribution problem.

---

† Here we assume that these initial temperatures do *not* bracket the melting point or other phase transition points at which there is a release of heat.

we have put a minus sign in the argument of the first erfc (4.13a). This is convenient for preserving the same form in both phases, and it again reflects the symmetry property of the heat equation under space inversion. The constants $A^{(1)}$ and $A^{(2)}$ are easily determined from the requirement of temperature and flux continuity† at the phase boundary:

$$T^{(1)} = T^{(2)}, \quad K^{(1)}\frac{\partial T^{(1)}}{\partial x} = K^{(2)}\frac{\partial T^{(2)}}{\partial x}$$

$$\text{for } x = 0\pm, \ t > 0. \tag{4.14a}$$

With Properties (a) and (c) of the last section, we get

$$A^{(1)} = \frac{T_\infty^{(2)} - T_\infty^{(1)}}{1 + (K^{(1)}/K^{(2)})(\kappa^{(2)}/\kappa^{(1)})^{1/2}}, \tag{4.14b}$$

$$A^{(2)} = \frac{T_\infty^{(1)} - T_\infty^{(2)}}{1 + (K^{(2)}/K^{(1)})(\kappa^{(1)}/\kappa^{(2)})^{1/2}}. \tag{4.14c}$$

Hence, we obtain the full temperature distributions (4.13) and the interface temperature

$$T_0 = \frac{K^{(1)}\sqrt{\kappa^{(2)}}\,T_\infty^{(1)} \dotplus K^{(2)}\sqrt{\kappa^{(1)}}\,T_\infty^{(2)}}{K^{(1)}\sqrt{\kappa^{(2)}} + K^{(2)}\sqrt{\kappa^{(1)}}}. \tag{4.15}$$

The interface temperature is thus a weighted mean of the initial temperatures in the two phases before contact. It is a *constant*, independent of time, for semi-infinite phases, and it is *not* a free parameter under the experimentalist's control. This is in line with the phase rule for isobaric, c-component, two-phase systems [*cf.* Eq. (B.3) of Appendix B]

$$f = c - 1 \quad (\varphi = 2, \ p = \text{const}), \tag{4.16}$$

---

† Flux continuity follows from energy conservation at *stationary* boundaries [*cf.* Eq. (1.19b)]. Temperature continuity, on the other hand, is really an additional assumption that can be understood as a condition of local thermal equilibrium [*cf.* Eq. (2.15a)]. There are cases where it may not hold.

where f is the number of degrees of freedom for such a system in equilibrium. In our case, the interface between the two different phases is assumed to be at equilibrium (*cf.* the last footnote), and it follows that there are no controllable parameters for one-component systems. For multicomponent systems, on the other hand, local equilibrium at the interface implies that we should really consider the redistribution of all species, as well as of energy.† It should be emphasized, however, that the interface temperature T_0 is a constant only for times such that the two adjacent phases can be considered semi-infinite. For finite phases and for long enough times, T_0's time dependence is controlled by the boundary conditions at the external phase boundaries.

Figure 4.5 displays the temperature distribution at various "times" $\sqrt{\kappa t}$, for the special case of equal thermal diffusivities and a ratio $K^{(2)}/K^{(1)} = 4$ of the thermal conductivities. The initial temperatures are taken to be $90°$C and $10°$C for phases (1) and (2), respectively, but any other values could be used by properly scaling these graphical results. The figure shows continuous and monotonically decreasing temperature profiles. At the interface there is an abrupt change in slope that corresponds to the conductivity change (*cf.* Exercise 1.4). The units for distance are determined by those of κ because, again, there are no natural, constant distance scales. As with all our previous examples, $\sqrt{\kappa t}$ is thus the penetration distance, which can be verified in Fig. 4.5.

Exercise 4.4: *Mass redistribution in a 1-d two-phase system*
Consider an isothermal, isobaric system of two large phases, such as Si/SiO$_2$, and the diffusion of an impurity whose constant levels $C_\infty^{(\alpha)}$ are initially given in both phases. Assume that the interface "partitions" the impurity, *i.e.*, the ratio $C^{(2)}/C^{(1)}$ at $x = 0$ is a known constant k (which can depend on temperature). Find the concentration distribution, and interpret the partitioning. Is the interface at "local thermodynamic equilibrium"? ∎

---

† In many cases one assumes "partial equilibrium" for those species that have diffusivities so low that they cannot redistribute significantly. Thus, one assumes that each phase *separately* is in chemical equilibrium, although together they are not.

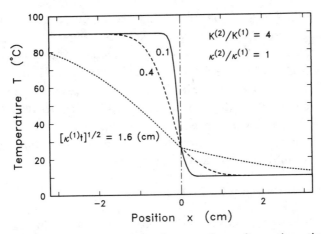

Fig. 4.5: Temperature distribution in the two phases for various times, measured in units of diffusion length.

A second interesting problem whose solution is expressed in terms of error functions is a simple 1-*d Stefan problem, i.e.,* the question of interface motion between two phases that support diffusion processes. For definiteness, consider the two-phase, binary system of Sects. 2.4-5, and assume that diffusion of component *B* can occur only in the liquid phase (2). According to the phase rule (4.16), there is but one degree of freedom at the interface if it is at local equilibrium. This is precisely what the phase diagrams in Figs. 2.5-6 tell us: The equilibrium com-

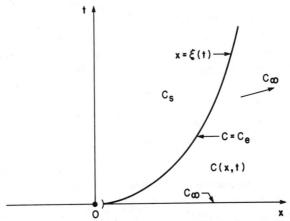

Fig. 4.6: Same as Fig. 4.4 for the one-phase Stefan problem.

positions of a (large) solid and liquid are uniquely determined at a given temperature. Take, now, a large solid of composition C_s, and put it in contact with a prepared semi-infinite liquid of *arbitrary* constant composition C_∞. What are we to expect? First, at the instant of contact, the surface concentration in the liquid must assume its equilibrium value C_e compatible with the phase diagram. Then, because generally $C_e \neq C_\infty$, we expect mass redistribution in the liquid, according to the diffusion equation (1.14), and consequent interface motion $\xi(t)$. The general solution to this problem is again

$$C(x,t) = C_\infty + A\,\mathrm{erfc}(x/2\sqrt{Dt}), \quad x > \xi(t), \tag{4.17}$$

because of the initial and boundary conditions shown in Fig. 4.6. In contrast to the first problem of this section, the interface can move, as was the case for the spherical precipitate of Sect. 2.5.† There are now two unknowns: the constant A in Eq. (4.17) and the interface's schedule $\xi(t)$. But we have, not so coincidentally, that many equations. These are the specification of local equilibrium

$$C(\xi,t) = C_e \tag{4.18a}$$

and of local mass conservation

$$\dot{\xi}(C_s - C_e) = DC_x \big|_{x=\xi(t)} \tag{4.18b}$$

at the moving boundary. [This last is the Stefan condition (1.20) if we neglect diffusion in the solid.] Inserting Eq. (4.17) into Eqs. (4.18), we get the requirements $C_e = C_\infty + A\,\mathrm{erfc}(\xi/2\sqrt{Dt})$ and $\dot{\xi} \propto t^{-1/2}\exp(-\xi^2/4Dt)$. These expressions cannot be satisfied unless $\xi \propto t^{1/2}$. Therefore, we have succeeded in finding the interface's motion — it is parabolic — which we write

$$\xi(t) = 2\lambda\sqrt{Dt}. \tag{4.19}$$

---

† We disallowed phase transitions in our previous thermal problem (4.12-14) and impurity diffusion problem (4.8), and there could be no interface motion. One can treat freezing and melting problems in the same way as the current constitutional problem in a binary [8,9]. Indeed, the freezing and melting of polar ice caps was Stefan's original concern.

The constant λ completely characterizes the interface's motion, both in magnitude and in sign. Positive values indicate growth and negative values indicate dissolution. We call it the *growth constant*. Then, inserting Eqs. (4.17) and (4.19) into the boundary conditions (4.18), we determine the constant

$$A = -\frac{C_\infty - C_e}{\mathrm{erfc}(\lambda)}, \tag{4.20a}$$

and we get the transcendental equation

$$\sqrt{\pi}\,\lambda e^{\lambda^2}\mathrm{erfc}(\lambda) = \frac{C_\infty - C_e}{C_s - C_e} \equiv \sigma \tag{4.20b}$$

for the growth constant λ. The right-hand side of Eq. (4.20b) is precisely the (absolute) supersaturation σ that appeared in Eqs. (2.25) for the spherical-precipitate problem. In both cases, σ represents the maximum driving force for the diffusion process. A given supersaturation determines λ according to Eq. (4.20b), hence the constant A through Eq. (4.20a), and thus the full concentration distribution (4.17).

Figure 4.7 shows a plot of the function on the left-hand side of Eq. (4.20b). Because of the asymptotic relations (4.11), that function has an upper bound for large positive values of λ, which correspond to

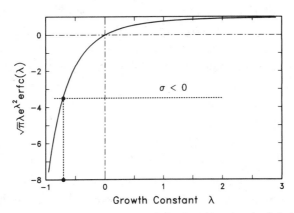

Fig. 4.7: The graph, as a function of λ, of the function on the left-hand side of Eq. (4.20b). This equation has a unique solution for each value of the absolute supersaturation σ (drawn negative on the figure).

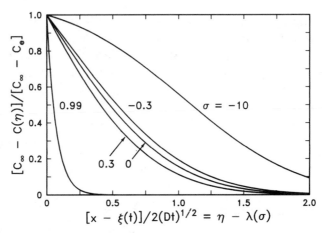

Fig. 4.8: Concentration distribution ahead of a moving interface, for various supersaturations.

$\sigma \rightarrow 1$. On the other hand, a Taylor expansion for small λ's shows that $\lambda \cong \sigma/\sqrt{\pi}$ when the supersaturation is small (*cf.* the next exercise). We note that there are no real solutions of Eq. (4.20b) for supersaturations greater than unity, but that case is excluded by the inequalities (2.19). This relation (4.20b) can be inverted graphically or numerically for $\lambda(\sigma)$, and Fig. 4.8 shows the concentration distribution that is obtained for various values of the supersaturation. The distribution (4.17) is defined on the right-hand domain of Fig. 4.6, *i.e.*, for $x \geq \xi(t)$. Anticipating somewhat the notion of similarity variable $\eta \equiv x/2\sqrt{Dt}$, explored in the next chapter, that condition reads simply $\eta \geq \lambda$. Note that positive σ's imply growth and concentration profiles without inflection, whereas negative σ's mean dissolution and the appearance of an inflection point in the concentration. Can you interpret this?

Exercise 4.5: *Behavior of the function* $\lambda(\sigma)$
Using analytic properties of error functions, expand the left-hand side of Eq. (4.20b) for small and large values of λ, both positive and negative. Thus, invert that equation in these limits. Verify the monotone behavior of the function in Fig. 4.7. ∎

There is much more to say on Stefan problems. They constitute an active area of research, probably because, in their full generality, they are very difficult. These problems are always nonlinear, although this may not yet be evident (see, however, Sect. 5.2). Some of the earlier work can be found in Refs. [8,9], and Rubinstein's treatise [10] is a classic investigation. Suffice it to say that there are very few analytical results (we will examine another in Sect. 5.3), and today's investigations, especially for multidimensional problems, are largely numerical [11-14].

4.4 THE GAUSSIAN HAS YET MORE RELATIVES

As mentioned at the beginning of the last section, all examples so far have imposed given conditions on the field at physical boundaries. Such a specification is known as a *Dirichlet condition*. There are evidently many other realistic types of boundary conditions. For example, consider the following *heating problem for a half space:* A large solid, initially at the constant temperature T_∞, is subjected to intense irradiation along its plane boundary, *e.g.*, by a laser or an electron beam.† By continuity (1.19), the imposed energy flux F must equal the thermal flux in the solid at that boundary:

$$F = -KT_x \big|_{x=0}.$$
(4.21a)

Since, as is easily verified, the flux $J = -KT_x$ satisfies the *same* heat equation (1.37) at all interior points of the solid as does temperature,‡ and since it satisfies the initial condition $J(x,0) = 0$, it follows that

$$J(x,t) = F \operatorname{erfc}(x/2\sqrt{\kappa t})$$
(4.21b)

gives the flux distribution. The temperature distribution is then ob-

---

† We assume that the source is not powerful enough to cause ablation: There are no moving boundaries in this problem. We also assume that the energy flux F is constant in time (more precisely, a step function). General time-dependent cases $F(t)$ are handled in Chap. 7, where we also explore the boundary condition (4.21a) in somewhat greater detail. The specification of the gradient at a boundary, rather than of the field, is called a *Neumann condition*.

‡ This is yet another manifestation of Property C.1 of Sect. 1.3.

tained by integrating this relation over space. Thus,

$$T(x,t) = (F/K) \int_x^\infty \mathrm{erfc}(x'/2\sqrt{\kappa t})\,dx' + f(t)$$

$$= (F/K)2\sqrt{\kappa t} \int_{x/2\sqrt{\kappa t}}^\infty \mathrm{erfc}(\eta)\,d\eta + T_\infty, \tag{4.21c}$$

where the arbitrary function of time $f(t)$ must reduce to the constant T_∞ because of the far-field condition on T. Consequently, the temperature distribution produced by a constant flux is given by the integral of an error function. Is this motivation enough to introduce the *iterated error functions*? In view of their wide application (laser processing being a simple example) we now turn to this family of functions.

Before doing so, however, we must introduce another useful special function, the *gamma function* and some of its properties [1-3]. Recall the standard result (4.2) and write that integral over $[0,\infty]$ in the new variable $\alpha = \xi^2$:

$$\sqrt{\pi} = \int_0^\infty \alpha^{-1/2} e^{-\alpha}\,d\alpha.$$

This integral is no more expressible in terms of elementary functions than is the integral (4.2) whence it came. When in doubt, give it a name $\Gamma(\tfrac{1}{2})$, and "imbed" this result in the following

Definition: The gamma function is defined by the integral

$$\Gamma(z) = \int_0^\infty \alpha^{z-1} e^{-\alpha}\,d\alpha. \tag{4.22}$$

This integral converges for positive values of z, and one shows that it diverges logarithmically when $z \to 0$. Integrating (4.22) by parts yields the recursion

$$\Gamma(z+1) = z\Gamma(z), \tag{4.23a}$$

and we have $\Gamma(1) = 1$ directly from the definition (4.22). Therefore, $\Gamma(2) = 1$, $\Gamma(3) = 2$,...., and generally, when z is a positive integer n, the gamma function interpolates the factorial

$$\Gamma(n+1) = n! \tag{4.23b}$$

of an integer. For this reason, Γ is sometimes called the factorial function. The recursion (4.23a) can be used in yet other ways. For example, with $\Gamma(\frac{1}{2}) = \sqrt{\pi}$, we have $\Gamma(3/2) = \Gamma(1+\frac{1}{2}) = \frac{1}{2}\Gamma(\frac{1}{2}) = \sqrt{\pi}/2$, and the recursion can be carried forward to any half-integer value $n+\frac{1}{2}$. Similarly, we use Eq. (4.23a) to continue the definition (4.22) to negative values of z [remember that the integral (4.22) converges only for positive values]. It suffices to replace z by $z-1$ in Eq. (4.23a), to get

$$\Gamma(z-1) = \frac{\Gamma(z)}{z-1}; \tag{4.23c}$$

hence, for example, $\Gamma(-\frac{1}{2}) = -2\sqrt{\pi}$. These results are generalized in the following

Exercise 4.6: *Gamma function for half-integer arguments*
Show that

$$\Gamma(n+\frac{1}{2}) = \frac{1\times3\times....\times(2n-1)}{2^n}\sqrt{\pi},$$

$$\Gamma(-n-\frac{1}{2}) = \frac{(-2)^{n+1}}{1\times3\times....\times(2n+1)}\sqrt{\pi}$$

for any non-negative integer n. Show, also, that the gamma function diverges for any negative integer: it has *poles* on the negative real axis. ∎

The gamma function enjoys many other interesting properties. Here, we mention only three that will prove useful. The first is a *duplication formula*, similar to what exists for trigonometric functions:

$$\sqrt{\pi}\,\Gamma(2z) = 2^{2z-1}\Gamma(z)\,\Gamma(z+\frac{1}{2}). \tag{4.24a}$$

The second is its asymptotic behavior for large positive z,

$$\Gamma(z+1) \sim \sqrt{2\pi z}\,(z/e)^z\left[1 + O(z^{-1})\right], \tag{4.24b}$$

also known as *Stirling's formula*. For proofs of these two properties (4.24) see Refs. [2,3]. The third property is the gamma function's ability to express some common integrals that involve exponentials. For example, by simple changes of variables and the definition (4.22), we easily get

$$\int_0^\infty \xi^{z-1} e^{-a\xi} d\xi = a^{-z} \Gamma(z), \tag{4.25a}$$

$$\int_0^\infty \xi^{2z-1} e^{-a\xi^2} d\xi = \frac{1}{2} a^{-z} \Gamma(z), \tag{4.25b}$$

for any positive values of the parameters z and a. In other words, arbitrary moments of exponentials and gaussians are expressible in terms of gamma functions. For arguments other than integers or half integers, the evaluation of the gamma function, like the error function, depends either on tables and rational-fraction approximations [1] or on the numerical evaluation of the integral (4.22).

We are now ready to attack the integrals of error functions.

Definition: The iterated error functions are defined by the recursion

$$i^n \text{erfc}(z) = \int_z^\infty i^{n-1} \text{erfc}(\xi) \, d\xi \qquad \forall n \geq 0, \; (n \text{ integer}), \tag{4.26a}$$

with the initial condition

$$i^{-1} \text{erfc}(z) = \frac{2}{\sqrt{\pi}} e^{-z^2}. \tag{4.26b}$$

With this definition, for example, the complementary error function is simply the zeroth member

$$i^0 \text{erfc}(z) = \text{erfc}(z) \tag{4.26c}$$

of the sequence (4.26a), and the solution to our previous heating problem (4.21c) is simply

$$T(x,t) = T_\infty + (F/K) 2\sqrt{\kappa t} \, i^1 \text{erfc}(x/2\sqrt{\kappa t}). \tag{4.21d}$$

Consequently, the surface temperature is

$$T(0,t) = T_\infty + (F/K) 2\sqrt{\kappa t} \, i^1 \text{erfc}(0). \tag{4.21e}$$

But aren't these merely empty definitions? And how do we compute these functions? The following properties show that iterated error functions are here to stay and that they are easy to evaluate.

a. Differentiation formula: From the definition (4.26a) we immediately get

$$\frac{d}{dz} i^n \text{erfc}(z) = -i^{n-1} \text{erfc}(z) \tag{4.27}$$

b. Expression as a single integral: According to the definition (4.26a), the nth iterated error function is the integral of the $(n-1)$th, and so forth, until we reach the gaussian (4.26b). We then have a sequence of nested integrals for which there is a theorem of calculus (*cf.* Sect. 6.5 of Chap. 6). Indeed, by repeated differentiation or by induction we verify that

$$i^n \text{erfc}(z) = \frac{2}{\sqrt{\pi}} \int_z^\infty \frac{(\xi - z)^n}{n!} e^{-\xi^2} d\xi. \tag{4.28}$$

c. Value at the origin: From this last result we compute

$$i^n \text{erfc}(0) = \frac{2}{\sqrt{\pi}} \int_0^\infty \frac{\xi^n}{n!} e^{-\xi^2} d\xi = \frac{1}{\sqrt{\pi}} \frac{1}{\Gamma(n+1)} \Gamma\left(\frac{n+1}{2}\right)$$

$$= \frac{1}{2^n \Gamma(1 + n/2)}, \tag{4.29}$$

where the second equality results from Eqs. (4.25b) and (4.23b), and where the duplication formula (4.24a) is used for the third equality. And, from Exercise 4.6, we know how to evaluate these gamma functions for integer n.

d. Recurrence relation: The iterated error functions obey an important recursion

$$2n \, i^n \text{erfc}(z) = -2z \, i^{n-1} \text{erfc}(z) + i^{n-2} \text{erfc}(z). \tag{4.30a}$$

It relates any member of that family of functions to its two immediate predecessors. For example, because of Eqs. (4.26b,c), we compute in succession

$$2i^1 \text{erfc}(z) = -2z \, \text{erfc}(z) + \frac{2}{\sqrt{\pi}} e^{-z^2}, \tag{4.30b}$$

$$4i^2 \text{erfc}(z) = (1 + 2z^2) \, \text{erfc}(z) - \frac{2z}{\sqrt{\pi}} e^{-z^2}. \tag{4.30c}$$

This allows us to evaluate these two functions because we know how to compute both the gaussian and the complementary error function. The proof of recursion (4.30a) is not hard: Just change variables $\xi \to \alpha = \xi - z$ in Eq. (4.28) and differentiate this result with respect to z, bearing the property (4.27) in mind. Figure 4.9 is a graph, for positive argument, of the first four iterated error functions. They all decrease monotonically with increasing argument, and they approach zero increasingly fast with the order n [cf. Eqs. (4.33) below].

e. Differential equation: Expressing the functions of lower order in (4.30a) as derivatives of the nth iterate, we find that $i^n \mathrm{erfc}(z)$ is a solution of the linear second-order differential equation (but with a variable coefficient of y')

$$y'' + 2zy' - 2ny = 0, \quad n \geq -1, \tag{4.31}$$

where primes mean differentiation with respect to z.

Exercise 4.7: *Another solution of this differential equation*
Show that $i^n \mathrm{erfc}(-z)$ is also a solution of the differential equation (4.31). Computing the Wronskian of the two functions $i^n \mathrm{erfc}(\pm z)$, show that they are linearly independent solutions of that equation if n is non-negative. ∎

f. Taylor series: Applying Taylor's theorem, together with the results (4.27) and (4.29), we get the expansion around the origin

$$i^n \mathrm{erfc}(z) = \sum_{k=0}^{\infty} \frac{z^k}{k!} \frac{d^k}{dz^k} i^n \mathrm{erfc}(z) \Big|_{z=0}$$

$$= \sum_{k=0}^{\infty} \frac{(-1)^k z^k}{k! \, 2^{n-k} \Gamma[1+(n-k)/2]}. \tag{4.32}$$

g. Asymptotic expansions: Under fairly general conditions, one shows [6] that asymptotic expansions can be integrated term by term. For example, integrating Eq. (4.11b) once by parts, we get

$$i^1 \mathrm{erfc}(z) \sim \frac{e^{-z^2}}{2z^2 \sqrt{\pi}} [1 + O(z^{-2})], \quad z > 0 \tag{4.33a}$$

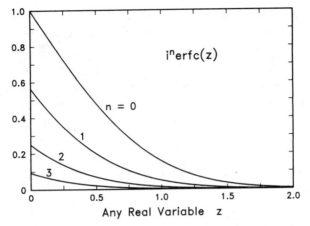

Fig. 4.9: The graph, for positive argument, of the first four iterated error functions.

and, more generally, one can show that

$$i^n\text{erfc}(z) \sim \frac{2}{\sqrt{\pi}} \frac{e^{-z^2}}{(2z)^{n+1}}\left[1 + O(z^{-2})\right] \tag{4.33b}$$

for large *positive* argument z.

Exercise 4.8: *Asymptotic behavior as $z \to -\infty$*
With the recursion (4.30a), show that

$$i^1\text{erfc}(z) \sim -2z \quad \text{and} \quad i^2\text{erfc}(z) \sim z^2 + \frac{1}{2}$$

for large *negative* argument (plus exponentially small terms). Can you generalize? ∎

   h. Solutions of the diffusion equation? By direct differentiation and substitution, we verify that $i^n\text{erfc}(x/2\sqrt{Dt})$ is *not* a solution of the diffusion equation. On the other hand, the function

$$C(x,t) \propto t^{n/2}\, i^n\text{erfc}(x/2\sqrt{Dt}), \quad n \geq -1, \tag{4.34}$$

is such a solution. (Please verify!) In fact, the solution (4.21d) of the previous thermal problem (4.21), that of Stokes's problem (4.9), and the gaussian (4.3) (the response to a point source) are precisely of this form. In these cases, we note that the surface fields (at $x = 0$) change as linear functions of $t^{1/2}$, t^0, and $t^{-1/2}$ respectively. In other words, the family of iterated error functions includes all our previous examples, and the functions (4.34) are therefore a natural extension of gaussian and error-function solutions to the diffusion equation. They describe diffusion when the surface concentration changes as an integer or half-integer power of time. We will take advantage of this property in the next section.

Exercise 4.9: *The gaussian generates the solutions (4.34)*
Consider a gaussian

$$G(x,t) = \frac{1}{\sqrt{\pi D t}} e^{-x^2/4Dt}$$

that is "twice as large" as the gaussian (4.3), *i.e.*, it is normalized in the interval $[0,\infty]$. Show that G generates the family (4.34) by repeated integration over space. ∎

4.5 CRYSTAL GROWTH UNDER CONDITIONS OF CONSTANT COOLING RATE

Our next application is of interest to those who grow crystals [15-17]. This problem may also be of interest to those who solve diffusion problems, because it introduces several new features: the possibility of decomposing a given problem into simpler parts (again, according to Property C.1 of Sect. 1.3) and the question of natural time scales. Consider the physical situation that we had encountered in Sects. 2.4-5 and in Sect. 4.3 above, namely, a binary liquid mixture brought into contact with a solid composed of the same elements. Here, we assume that the crystal is large in extent, and we focus our attention on a particular crystal face. Diffusion in the liquid, perpendicular to that face, is then essentially one-dimensional. Further, as we had noted, the liquid could be arbitrarily prepared, *i.e.*, its composition C_∞ in the minor component B is in the hands of the experimentalist. But he can do more than just join these two phases and let the concentration relax towards equilibrium. He can also impose controlled temperature variations on his solid-liquid system, *e.g.*, by slowly ramping the power of his furnaces. This creates an extra driving force for diffusion because the equilibrium concentration shifts with temperature, as permitted by the phase rule (4.16): For binary, isobaric, two-phase systems, there is a single degree of freedom, which can be used to advantage, for example, by ramping the temperature according to the schedule

$$T(t) = T_0 - \alpha t, \tag{4.35a}$$

where T_0 is the initial temperature and α is the cooling rate.† Since the liquidus (cf. Fig. 2.6) is described by some $C_e(T)$ relation, it follows that the equilibrium concentration changes with time according to

$$C_e(T) \cong C_e(T_0) + (T - T_0)\frac{dC_e}{dT}\Big|_{T=T_0} + O[(T-T_0)^2]$$

$$= C_0(1 - \alpha t/mC_0) + O(t^2). \tag{4.35b}$$

In this second equality we have used Eq. (4.35a), and we define C_0 and m as the equilibrium concentration $C_e(T_0)$ and liquidus slope $dT/dC_e|_{T_0}$ at the initial temperature, respectively.

An exact solution of this Stefan problem is unavailable, but if we assume that the crystal-liquid interface moves slowly enough, then, as in the spherical-precipitate problem of Sect. 2.5, we can solve the diffusion problem as if that interface were at rest. We then use the Stefan condition (1.20) to determine its motion. The diffusion problem is quite simple: The concentration in the liquid must be equal, initially and at infinity, to its prepared value C_∞, whereas, if the interface is again assumed to be at local thermodynamic equilibrium, then the surface concentration must be equal to that determined by Eq. (4.35b). Mathematically, this means

$$C(x,0) = C(\infty,t) = C_\infty, \tag{4.36a}$$

---

† Thermal diffusivities are generally much larger than mass diffusivities in liquids and solids. Therefore, temperature equilibration occurs must more quickly than mass redistribution, and we can assume that the temperature is spatially constant. The temperature ramp rate α in Eq. (4.35a) is, here, defined positive for *cooling*. Negative α's correspond to heating.

$$C(0,t) = C_0(1 - \alpha t/mC_0) \tag{4.36b}$$

if the interface position essentially coincides with the spatial frame's origin. The extra twist here is the surface concentration's *given* (linear) time dependence. This Eq. (4.36b) deserves yet another remark. Although the temperature schedule (4.35a) has a characteristic time constant $T_0/|\alpha|$, the system responds with a time constant

$$\tau = mC_0/|\alpha|, \tag{4.35c}$$

and these need not be at all equal.

The solution to this diffusion problem follows easily if we recognize that the concentration can be viewed as the superposition $C = C_1 - C_2$ of two partial concentrations, each of which satisfies the diffusion equation under the conditions

$$C_1(x,0) = C_1(\infty,t) = C_\infty, \tag{4.37a}$$

$$C_1(0,t) = C_0, \tag{4.37b}$$

and

$$C_2(x,0) = C_2(\infty,t) = 0, \tag{4.38a}$$

$$C_2(0,t) = \text{sgn}(\alpha)\, C_0 t/\tau. \tag{4.38b}$$

The solution for C_1 has been derived a number of times in this chapter. Since its initial and boundary conditions (4.37) are constant, we have

$$C_1(x,t) = C_\infty - (C_\infty - C_0)\,\text{erfc}(x/2\sqrt{Dt}). \tag{4.39}$$

On the other hand, the partial concentration C_2 must vanish initially and at infinity, and it must be proportional to the first power of time at the interface. Looking back at our lexicon of available solutions (4.34) in unbounded media, we see that the second iterated error function fits our conditions, and we tentatively write $C_2 = A t\, \mathrm{i}^2\text{erfc}(x/2\sqrt{Dt})$, where the constant A follows by matching this solution at the boundary with condition (4.38b). This solution satisfies the other conditions (4.38a) by virtue of the asymptotic expansion (4.33b). But $i^2 \text{erfc}(0) = 1/4$, in accordance with Eqs. (4.29) and (4.23b), so that $A = \text{sgn}(\alpha)\, 4C_0/\tau$, and we get

$$C_2(x,t) = \text{sgn}(\alpha)\, (4C_0 t/\tau)\, i^2 \text{erfc}(x/2\sqrt{Dt}). \tag{4.40}$$

To get the growth rate and hence the interface's schedule, we need only remember that the Stefan condition (1.20) must hold. (It is again assumed that diffusion in the solid phase is negligible.) Therefore, using the differentiation formula (4.27), we get

$$\dot{\xi}\,[C_s - C(0,t)] = D \frac{\partial C}{\partial x}\Big|_{x=0} = D\left(\frac{\partial C_1}{\partial x}\Big|_{x=0} - \frac{\partial C_2}{\partial x}\Big|_{x=0} \right)$$

$$= \frac{1}{\sqrt{\pi}}(C_\infty - C_0)(D/t)^{1/2} + \text{sgn}(\alpha)(2C_0/\tau)(Dt/\pi)^{1/2}. \tag{4.41a}$$

With Eq. (4.36b), this is a quadrature for the interface schedule $\xi(t)$. The integration is notably simpler if we only consider times short with respect to the relaxation time τ. Then the surface concentration remains essentially at its initial value C_0, and we get the dimensionless form

$$\frac{\xi(t)}{\sqrt{D\tau}} \cong \frac{2}{\sqrt{\pi}} \frac{C_\infty - C_0}{C_s - C_0}(t/\tau)^{1/2}$$

$$+ \text{sgn}(\alpha) \frac{4}{3\sqrt{\pi}} \frac{C_0}{C_s - C_0}(t/\tau)^{3/2}. \tag{4.41b}$$

This expression is the sum of two terms that correspond to the two driving forces of our system. The first, parabolic and proportional to $t^{1/2}$, is caused by the initial (absolute) supersaturation σ in the liquid. It is no different from the many other diffusion-limited examples we have seen. On the other hand, the second term is proportional to $t^{3/2}$; it is due to the *additional* supersaturation caused by temperature ramping. This second driving force is caused by agencies *exterior* to our system. Both terms can have either sign, and, thus, regimes of crossover from growth to dissolution can occur. Figure 4.10 shows just such a case. The function (4.41b) is plotted for the case of cooling, $\alpha > 0$, and for various values of the relative supersaturation $\sigma_r = (C_\infty - C_0)/C_0$. It is

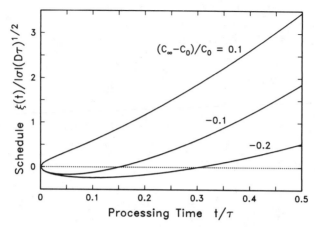

Fig. 4.10: Interface schedule for a crystallizing system under two distinct driving forces: initial supersaturation and cooling. Note the period of dissolution when these forces are opposed.

evident, when the system is initially undersaturated, that some time must elapse before the crystal can overcome this condition and begin to grow. This is sometimes observed and is called, somewhat in desperation, an "incubation time." The crossover from dissolution to growth occurs at a time

$$t = -\text{sgn}(\alpha)\, 3\sigma_r \tau/2 \tag{4.41c}$$

if the ramp rate and the relative supersaturation have opposite sign. On the other hand, the crystal always grows (or dissolves) monotonically when α and σ_r have the same sign.

The problem which we have just solved suggests an immediate generalization. If, for example, the surface concentration is given as a *known* power series in time, then, by virtue of the basic solutions (4.34), we expect a power-series solution in the *even* iterated error functions. In fact, we shall see in Eq. (6.36) of Chap. 6 that there exists a closed-form representation of the diffusion equation's solution if the surface concentration is an *arbitrarily given* function of time. Likewise, the last chapter will address the general problem of thermal stimulation, *i.e.*, situations where temperature programming can be used to advantage to study diffusion and desorption. The procedure of this section is amplified in the next

Exercise 4.10: *Solutions in a half space for variable Dirichlet conditions*

Assume that the surface concentration is a known function of time and that it can be expanded as a Taylor series in $\sqrt{t}$:

$$C_0(t) = \sum_{n=0}^{\infty} a_n t^{n/2}.$$

In addition, assume constant initial and far-field conditions such as given by Eqs. (4.36a). Find the solution of the diffusion equation. What can you say about the surface flux? Can you generalize when the series for C_0 also contains a finite number of terms with negative powers $n \leq -1$? (This is called a Laurent series.) Can you solve similar problems in which the applied flux F, rather than C_0, is a known Laurent series in $\sqrt{t}$? ∎

4.6 PRODUCT SOLUTIONS FOR HIGHER-DIMENSIONAL PROBLEMS

We have just seen that linear combinations (*i.e.*, essentially *sums*) of the diffusion equation's solutions can help solve problems when the boundary conditions themselves are linear combinations of powers of time. We now ask whether *products* of solutions can be put to advantage. For simplicity, we illustrate the procedure for two dimensions and in cartesian coordinates (x,y). Let $C_1(x,t)$ and $C_2(y,t)$ be any solutions of the 1-*d* diffusion equations

$$\frac{\partial C_1}{\partial t} = D\frac{\partial^2 C_1}{\partial x^2} \quad \text{and} \quad \frac{\partial C_2}{\partial t} = D\frac{\partial^2 C_2}{\partial y^2} \qquad (4.42a,b)$$

respectively. For the moment, we specify neither the domains† of the independent variables (x,t) and (y,t) nor the initial and boundary conditions. Nevertheless, the product

$$C(x,y,t) \equiv C_1(x,t)\,C_2(y,t) \qquad (4.43)$$

---

† Topologically, a *domain* is any open and connected set. Here, we deal with ordered pairs or triples (enclosed in parentheses) of real numbers. Curly braces containing inequalities on the independent variables define domains. The reader is advised to sketch, as we go along, the indicated domains and the initial and boundary conditions.

satisfies the 2-d equation

$$\frac{\partial C}{\partial t} = D\left(\frac{\partial^2 C}{\partial x^2} + \frac{\partial^2 C}{\partial y^2} \right) \tag{4.44}$$

over some domain of the variables (x,y,t), because its substitution into Eq. (4.44) yields

$$C_2 \frac{\partial C_1}{\partial t} + C_1 \frac{\partial C_2}{\partial t} = D\left(C_2 \frac{\partial^2 C_1}{\partial x^2} + C_1 \frac{\partial^2 C_2}{\partial y^2} \right), \tag{4.45}$$

and this is an identity by virtue of Eqs. (4.42).

For example, the functions

$$C_1 = \text{erfc}(x/\sqrt{Dt}) \quad \text{and} \quad C_2 = \text{erfc}(y/\sqrt{Dt}) \tag{4.46a,b}$$

are solutions of Eqs. (4.42) over the domains

$$\{x > 0, t > 0\} \quad \text{and} \quad \{y > 0, t > 0\}, \tag{4.47a,b}$$

respectively, and they satisfy the initial and boundary conditions

$$C_1(x,0) = C_1(\infty,t) = 0, \quad C_1(0,t) = 1, \tag{4.48a}$$

$$C_2(y,0) = C_2(\infty,t) = 0, \quad C_2(0,t) = 1. \tag{4.48b}$$

Consequently, their product $C = \text{erfc}(x/\sqrt{Dt})\,\text{erfc}(y/\sqrt{Dt})$, defined over the domain $\{x > 0, y > 0, t > 0\}$, satisfies Eq. (4.44), the initial condition $C(x,y,0) = 0$, and the boundary conditions at infinity $C(x,\infty,t) = C(\infty,y,t) = 0$, all potentially reasonable requirements. But the boundary conditions at $x = 0$ and $y = 0$ would be most peculiar because, by construction, we have $C(0,y,t) = \text{erfc}(y/\sqrt{Dt})$ and $C(x,0,t) = \text{erfc}(x/\sqrt{Dt})$, which are unlikely conditions in practice. This product solution is therefore *not* a suitable candidate, but it does indicate that we require *homogeneous* conditions† at finite boundaries. For example, the product

$$C(x,y,t) = \text{erf}(x/\sqrt{Dt})\,\text{erf}(y/\sqrt{Dt}), \tag{4.49a}$$

---

† A homogeneous boundary condition is any relation of the form $f(C, \nabla C) = 0$ that vanishes with both C and ∇C. For example, $\alpha C + \beta\, \partial C/\partial x = 0$ is homogeneous.

defined on the same "octant," satisfies the initial and boundary conditions

$$C(x,y,0) = C(\infty,y,t) = C(x,\infty,t) = 1, \qquad (4.49b)$$

$$C(0,y,t) = C(x,0,t) = 0. \qquad (4.49c)$$

These correspond to a concentration distribution about a rectangular corner. The initial concentration is everywhere unity, and the edges of the corner are held at zero. Evidently, we can multiply this solution by arbitrary constants and add other arbitrary constants to produce new solutions, where, for example, the concentration on the corner's edges could be a given constant.

Exercise 4.11: *The case of arbitrary initial conditions*
Consider solutions $C_1(x,t)$ and $C_2(y,t)$ of equations (4.42) that satisfy *any* homogeneous conditions on finite boundaries and that go to the same constant at infinity if the domain is unbounded. Let $\varphi_1(x)$ and $\varphi_2(y)$ be their initial conditions. Show that the function C, defined as the product (4.43), is a solution of Eq. (4.44) over the octant $\{x > 0, y > 0, t > 0\}$ if it satisfies the same homogeneous boundary conditions and if its initial condition $\varphi(x,y)$ is precisely $\varphi_1 \varphi_2$. ∎

This method of construction, it appears, requires considerable skill. It is explained in more detail on pp. 33-35 of Carslaw and Jaeger's exhaustive survey [8] of known solutions, and it is extensively used in their Chaps. 5-7. Today, one tends to handle multidimensional problems through repeated Fourier transforms [18].

We end this section with a few remarks about the method of *separation of variables*. As is well known, solutions of partial differential equations are often sought in the form of products. What relation, if any, is there with the previous method? To illustrate, take again the case of Eq. (4.44), and assume that its solutions C can be expressed as products of functions involving fewer variables. For example, assume that

$$C(x,y,t) = X(x,t) Y(y,t), \qquad (4.50)$$

where X and Y are functions of the indicated variables. Introducing this product into Eq. (4.44), we also get the form (4.45), which we write

$$\frac{1}{X}\left[\frac{\partial X}{\partial t} - D\frac{\partial^2 X}{\partial x^2}\right] = -\frac{1}{Y}\left[\frac{\partial Y}{\partial t} - D\frac{\partial^2 Y}{\partial y^2}\right]. \tag{4.51}$$

Now, however, the functions X and Y are *not* necessarily solutions of Eqs. (4.42). Consequently, Eq. (4.51) is not an identity. Nevertheless, because the left- and right-hand sides of this Eq. (4.51) are functions of different sets of variables, it follows that each side must be separately equal to the same arbitrary function λ of the common variable t. (Can you prove this?) Consequently, we have separated the 2-d equation (4.44) into the two 1-d equations

$$\frac{\partial X}{\partial t} = D\frac{\partial^2 X}{\partial y^2} + \lambda(t)X, \tag{4.52a}$$

$$\frac{\partial Y}{\partial t} = D\frac{\partial^2 Y}{\partial y^2} - \lambda(t)Y, \tag{4.52b}$$

and the arbitrary function λ is yet to be determined from the initial and boundary conditions on C. Equations (4.52) are diffusion equations with sources.

REFERENCES

1. M. Abramowitz and I.A. Stegun, *Handbook of Mathematical Functions* (reprinted by Dover, New York 1965).

2. N.N. Lebedev, *Special Functions and their Applications,* R.A. Silverman, Ed. (reprinted by Dover, New York 1972).

3. D.V. Widder, *Advanced Calculus,* 2nd ed. (Prentice-Hall, Englewood Cliffs, NJ 1961).

4. B. Gnedenko, *The Theory of Probability* (Mir Publishers, Moscow 1976).

5. R. Ghez, G.S. Oehrlein, T.O. Sedgwick, F.F. Morehead, and Y.H. Lee, "Exact Description and Data-Fitting of Ion-Implanted Dopant Profile Evolution during Annealing," *Appl. Phys. Lett.* **45**, 881 (1984).

6. A. Erdélyi, *Asymptotic Expansions* (reprinted by Dover, New York 1956).

7. C.C. Lin and L.A. Segel, *Mathematics Applied to Deterministic Problems in the Natural Sciences* (MacMillan, New York 1974).

8. H.S. Carslaw and J.C. Jaeger, *Conduction of Heat in Solids,* 2nd ed., pp. 282-296 (Oxford University Press, London 1959).

9. J. Crank, *The Mathematics of Diffusion,* 2nd ed., pp. 286-325 (Oxford University Press, London 1975).

10. L.I. Rubinstein, *The Stefan Problem* (Translations of Mathematical Monographs, Vol. 27, by American Mathematical Society, Providence, RI 1971).

11. J.M. Ockendon and W.R. Hodgkins, Eds., *Moving Boundary Problems in Heat Flow and Diffusion* (Oxford University Press, London 1975).

12. G.H. Meyer, "An Alternating Direction Method for Multi-dimensional Parabolic Free Surface Problems," *Int. J. Numer. Methods Engng.* **11**, 741 (1977).

13. R.S. Gupta and D. Kumar, "Variable Time Step Methods for One-dimensional Stefan Problem with Mixed Boundary Conditions," *Int. J. Heat Mass Transf.* **24**, 251 (1981).

14. J.J. Derby and R.A. Brown, "A Fully-implicit Method for Simulation of the One-dimensional Solidification of a Binary Alloy," *Chem. Engng. Sci.* **41**, 37 (1986).

15. M.B. Small and J.F. Barnes, "The Distribution of Solvent in an Unstirred Melt under the Conditions of Crystal Growth by Liquid Epitaxy and its Effect on the Rate of Growth," *J. Cryst. Growth* **5**, 9 (1969).

16. E.A. Giess and R. Ghez, "Liquid-Phase Epitaxy," in *Epitaxial Growth,* J.W. Matthews, Ed., Part A, pp. 183-213 (Academic Press, New York 1975).

17. M.B. Small and R. Ghez, "Conditions for Constant Growth Rate by LPE from a Cooling, Static Solution," *J. Cryst. Growth* **43**, 512 (1978). [There is a small misprint in Eq. (5) of this paper; its right-hand side should read $(4/3\sqrt{\pi})(t/\tau)^{3/2}$.]

18. R.V. Churchill, *Operational Mathematics*, 3rd ed. (McGraw-Hill, New York 1972).

Chapter 5

AN INTRODUCTION TO SIMILARITY

With iterated error functions we have already greatly expanded our vocabulary of solutions of the diffusion equation. The question is, however, whether or not any useful further expansion should be expected. The study of physical symmetries provides a partial answer, and that study is called similarity. In essence, we shall find that the linear diffusion equation in 1-d can have two types of solutions: superpositions either of iterated error functions or of trigonometric functions, otherwise known as Fourier series. The strength of similarity methods lies in their applicability to nonlinear problems as well. This chapter continues with more physical examples and with the construction of their solutions. These include exact solutions for the kinetics of phase transformations and for those examples where the diffusivity depends on the field. In contrast, the next chapters describe deductive methods for the solution of diffusion problems.

5.1 SIMILARITY SOLUTIONS

In Geometry, two figures are said to be similar if they are congruent after a uniform stretching of space. In turn, congruent figures are those which can be superposed after a rigid-body motion (rotations and translations, and, perhaps, reflections of space). The key notion for geometric similarity is our ability to change the length scale and, so doing, to introduce an equivalence between certain geometric figures. Theorems about right triangles, say, hold for *any* member of that class, rather than for this or that particular triangle. Similarity is also a useful tool to investigate classes of solutions of differential equations. In our case, the diffusion equation, we now ask if there exists a real *scale function* $T(t)$ of the dependent variable C and another, $L(t)$, of the independent variable x such that, for any two times t_1 and t_2, there exist two locations x_1 and x_2 where the solution "looks alike." In other words, when distance x is properly scaled, the ratios

$$\frac{C(x_1, t_1)}{T(t_1)} = \frac{C(x_2, t_2)}{T(t_2)} \tag{5.1a}$$

must have the same numerical value for all pairs of values of time t_1 and t_2. This is evidently possible if the concentration has the form

$$C(x,t) = T(t)\, F(\eta), \quad \text{with} \quad \eta = x/L(t). \tag{5.1b}$$

In that case the solution is the product of a pure function T of time with a function F that depends, at most, on a so-called *similarity variable* η. Such a solution (5.1b) is called a *similarity solution*. The search for these solutions is an important aspect of hydrodynamics [1,2] because of the Navier-Stokes equations' complexity: they are nonlinear. That search is likewise instructive for the diffusion equation, even though that equation is linear when the diffusivity is independent of the field.†

Before proceeding with the formal development, let us note that we have already encountered similarity in at least two guises. The error-function solutions (4.8), (4.9), (4.13), and (4.17) of particular physical problems were all similarity solutions. In these cases, the concentration scale T was a constant, and the length scale was simply $L = 2\sqrt{Dt}$. Likewise, the class of solutions (4.34) [and the related physical problems (4.3), (4.21), and (4.40)] was of the form (5.1b) if T is a power of $t^{1/2}$. Our other encounter with similarity was through Exercises 1.5 and 1.13. There, we found that the diffusion equation was invariant under the change of variables

$$\bar{x} = \alpha x + x_0 \quad \text{and} \quad \bar{t} = \alpha^2 t + t_0. \tag{5.2}$$

It follows that $\bar{x}/\sqrt{\bar{t}} = x/\sqrt{t}$ when the translations x_0 and t_0 are zero. In other words, the scale function $L \propto \sqrt{t}$ could have been predicted for the physical problems just mentioned. But are there any other possible

---

† Even the linear diffusion equation (1.14) can lead to a nonlinear problem because nonlinearities can arise from the boundary conditions. We shall have occasion to meet such examples in the next section and in Chap. 7.

scale functions? A second prediction follows from the transformation (5.2) by putting $\alpha = 1$ and the origin of time at $\bar{t} = 0$. We are then left with pure translations of space $\bar{x} = x + x_0$ and $\bar{t} = t$, which means that solutions can be periodic in space with period x_0. What is the meaning of periodicity, and what class of functions must be considered? We now turn to the consequences of similarity (5.1).

We wish to find the scales T and L such that the function (5.1b) is a solution of the diffusion equation (1.14). Simply compute the partial derivatives

$$\frac{\partial C}{\partial t} = \dot{T}F + T\left(-\frac{x\dot{L}}{L^2}\right)F', \tag{5.3a}$$

$$\frac{\partial C}{\partial x} = \frac{T}{L}F' \quad \text{and} \quad \frac{\partial^2 C}{\partial x^2} = \frac{T}{L^2}F'', \tag{5.3b}$$

where the dots and primes indicate ordinary derivatives with respect to t (for T and L) and to η (for F), respectively. Inserting the partial derivatives (5.3) into the diffusion equation, we get

$$DF'' + (L\dot{L})\eta F' - (\dot{T}L^2/T) F = 0, \tag{5.3c}$$

an expression we would like as simple as possible. Viewed as a differential equation for the function $F(\eta)$, Eq. (5.3c) reduces to an *ordinary* differential equation in $F(\eta)$ if the coefficients in parentheses are independent of time. Thus, we impose the subsidiary conditions

$$L\dot{L} = 2D\alpha, \tag{5.4a}$$

$$\dot{T}L^2/T = D\beta, \tag{5.4b}$$

where α and β are constants, and the factor D is thrown in for convenience. Equation (5.3c) then becomes simply

$$F'' + 2\alpha\eta F' - \beta F = 0. \tag{5.4c}$$

The condition (5.4a) is easily integrated with respect to t

$$L^2(t) = L_0^2 + 4\alpha Dt, \tag{5.5a}$$

where L_0 is the initial value $L(0)$. Inserting this function into the second condition (5.4b), we get

$$T(t) = \gamma(L_0^2 + 4\alpha Dt)^{\beta/4\alpha}. \tag{5.5b}$$

Having found the functional form of the scales, these are now discussed in terms of the four integration constants: α, β, L_0, and γ. By "discussion" is meant the search for *physical solutions*, *i.e.*, as in mechanics, those solutions that are *real* and that are *bounded* except, perhaps, at isolated points in the (x,t)-plane.

We first observe that the multiplicative constant γ can be chosen arbitrarily because the (linear) diffusion equation is homogeneous in C, and thus in T. Then, α must be real and non-negative, or else there would be times such that L, given by Eq. (5.5a), became imaginary. Accordingly, we distinguish the two cases: $\alpha = 0$ and $\alpha > 0$.

Case $\alpha = 0$: In this case L_0 must be nonzero, or else there would be no length scale (5.5a) at all. Choose the value $|L_0|^{-\beta/2\alpha}$ for the arbitrary constant γ, so that Eq. (5.5b) yields

$$T = \lim_{\alpha \to 0} (1 + 4\alpha Dt/L_0^2)^{\beta/4\alpha} = e^{\beta Dt/L_0^2}. \tag{5.6a}$$

On the other hand, Eq. (5.4c) becomes simply

$$F'' - \beta F = 0 \tag{5.6b}$$

in the similarity variable

$$\eta = x/L = x/(\pm |L_0|). \tag{5.6c}†$$

As just mentioned, we seek bounded solutions, in particular, those which decay for long times. It follows that the exponent in Eq. (5.6a) must be negative, ‡ and thus $\beta = -\lambda^2$. The solutions of Eq. (5.6b) are then trigonometric functions, and, collecting our partial results, we have

---

† The double sign again reflects the symmetry of the diffusion equation with respect to space inversion.

‡ Strictly speaking, the exponent must have negative *real* part. Some physical problems, such as the diurnal temperature variations of the earth, are periodic in time. These problems require a complex exponent, yet also with negative real part.

$$C \propto e^{-\lambda^2 Dt/L_0^2} e^{\pm i\lambda x/|L_0|}. \tag{5.7}$$

But this is merely the solution that is usually obtained under the heading of method of *separation of variables*. This method, usually introduced without much justification, now appears as a special case of similarity. The constant length scale L_0, in general, is determined by natural length units of a given physical problem. Thus, we anticipate that solutions of the form (5.7) apply mainly to problems in *bounded* regions.

Case $\alpha > 0$: In this case we can put $L_0 = 0$ by an appropriate choice of the arbitrary origin of time in Eq. (5.5a). Then, for convenience, choose $\gamma = \alpha^{-\beta/4\alpha}$ for the arbitrary integration constant in Eq. (5.5b). It follows that the scale functions are now

$$L(t) = 2(\alpha Dt)^{1/2}, \tag{5.8a}$$

$$T(t) = (4Dt)^{\beta/4\alpha}. \tag{5.8b}$$

Next, note that the transformation $\bar{\eta} = \eta\sqrt{\alpha} = x/2\sqrt{Dt}$ allows us to express the differential equation (5.4c) as

$$\frac{d^2 F}{d\bar{\eta}^2} + 2\bar{\eta}\frac{dF}{d\bar{\eta}} - \frac{\beta}{\alpha}F = 0. \tag{5.8c}$$

We now state, without proof (*cf.* [3, p. 293]), that the only bounded solutions of this differential equation are those for which the constant β/α is an even integer $2n$.† But then what do we have but the differential equation Eq. (4.31) for the iterated error functions! Collecting our results (5.8), we obtain the similarity solutions (4.34)

$$C \propto (4Dt)^{n/2} i^n \text{erfc}(\pm x/2\sqrt{Dt}) \qquad \forall \text{ integer } n \geq -1 \tag{5.9}$$

that were derived earlier by construction, and *they are the only ones* when $\alpha > 0$. Since here there is no constant length scale, but rather a

---

† The integer n can have either sign, so that, strictly speaking, the class of allowable functions is larger than (5.9): It includes Hermite polynomials multiplied by gaussians. These functions, however, are rarely used for diffusion, although they are essential for the analysis of the quantum-mechanical harmonic oscillator.

diffusion length $L \propto \sqrt{t}$ that increases indefinitely with time, it follows that these solutions apply mainly to problems in *unbounded* regions.

In sum, similarity imposes strict conditions on the possible physical solutions of the diffusion equation. They must have either the form (5.7) or the form (5.9), possibly augmented by translations of the independent variables. Both forms result from the solution of ordinary differential equations. These will be applied to specific physical problems in the next sections.

We end this section on a note of hope for the *nonlinear* diffusion equation (1.15). Indeed, if the diffusivity varies with concentration, then sometimes there also exist similarity solutions that stem from the solution of ordinary differential equations. Some of these cases will be taken up late in this chapter. A careful study, however, requires the theory of Lie groups. This is beyond our scope, and we merely point the reader to some of the relevant literature [4-6]. Nonetheless, like Molière's *Monsieur Jourdain*, in this section we have, in fact, just analyzed — prosaically — two particular subgroups of the diffusion equation's group of physical symmetries.

5.2 BOLTZMANN'S TRANSFORMATION

It is perhaps of more than historical interest that the question of similarity for diffusion in unbounded phases was first raised by Boltzmann [7]. With remarkable intuition, he introduced the notion of similarity variable $\eta = x/2\sqrt{Dt}$ for the solution of a problem in nonlinear diffusion. A transformation of the diffusion equation only in terms of that variable is called *Boltzmann's transformation*. It one of the more fruitful ideas for the solution of certain nonlinear problems. For example, it is instructive to rework our previous Stefan problem (4.17-20) in this light. This problem is nonlinear, as will be shown presently, even for cases of constant diffusivity. In fact, as mentioned earlier, almost all moving-boundary problems are nonlinear.

Fig. 5.1: Schematic representation of a frame moving with the phase boundary.

Recall, first, that we "guessed" the similarity solution (4.17) in the laboratory frame (x,t).† In that frame, the solid-liquid interface moves with a schedule $\xi(t)$ given by Eq. (4.19). Rather than guess, we now change coordinates to an accelerated frame that "immobilizes" the boundary [8]:

$$\bar{x} = x - \xi(t) \quad \text{and} \quad \bar{t} = t. \tag{5.10a}$$

In other words, the unfortunate observer of Fig. 5.1, rigidly skewered on the $(\bar{x},\bar{t})$-frame, would always measure $\bar{x} = 0$ for the interface's position. Moreover, this frame is accelerated because ξ can be an *a priori arbitrary* function of time. The formulas (5.10a) thus constitute an extension of galilean transformations that were included in the

---

† By "laboratory frame" we mean a spatial reference frame in which the the diffusion equation (1.14) or (1.15) is valid, *e.g.*, a frame rigidly attached to the solid. This is not quite accurate for the case at hand if solid and liquid have different densities, for then the liquid has a convective motion induced by the interface's motion (*e.g.*, see Ref. [24] of Chap. 2 or [11] of this chapter).

affine transformations of Exercises 1.5 and 1.13. As in these exercises, we must compute partial derivatives with respect to the new variables:

$$\frac{\partial C}{\partial x} = \frac{\partial C}{\partial \bar{x}}\frac{\partial \bar{x}}{\partial x} + \frac{\partial C}{\partial \bar{t}}\frac{\partial \bar{t}}{\partial x} = \frac{\partial C}{\partial \bar{x}}, \tag{5.10b}$$

$$\frac{\partial C}{\partial t} = \frac{\partial C}{\partial \bar{x}}\frac{\partial \bar{x}}{\partial t} + \frac{\partial C}{\partial \bar{t}}\frac{\partial \bar{t}}{\partial t} = \frac{\partial C}{\partial \bar{x}}(-\dot{\xi}) + \frac{\partial C}{\partial \bar{t}}. \tag{5.10c}$$

If these are inserted into the Stefan condition (4.18b) and the diffusion equation (1.14), then we get

$$\dot{\xi}(C_s - C_e) = DC_{\bar{x}}\big|_{\bar{x}=0}, \tag{5.11a}$$

$$C_{\bar{t}} = DC_{\bar{x}\bar{x}} + \dot{\xi}C_{\bar{x}}. \tag{5.11b}$$

Clearly, we have bought a simpler Stefan condition† at the expense of a convective term $\dot{\xi}C_{\bar{x}}$ in the diffusion equation (5.11b). This term is nonlinear in C because, eliminating $\dot{\xi}$ with Eq. (5.11a), it is proportional to $C_{\bar{x}}(0,\bar{t})C_{\bar{x}}(\bar{x},\bar{t})$. Following Boltzmann, let us *assume* that the concentration distribution depends only on the similarity variable

$$\bar{\eta} = \bar{x}/2\sqrt{D\bar{t}}. \tag{5.12a}$$

Then we must compute the partial derivatives

$$\frac{\partial C}{\partial \bar{x}} = \frac{\partial C}{\partial \bar{\eta}}\frac{1}{2\sqrt{D\bar{t}}} \quad \text{and} \quad \frac{\partial C}{\partial \bar{t}} = \frac{\partial C}{\partial \bar{\eta}}\left(-\frac{\bar{\eta}}{2\bar{t}}\right). \tag{5.12b}$$

Inserting these into Eqs. (5.11) and clearing fractions, we get

$$2\dot{\xi}\sqrt{\bar{t}/D}\,(C_s - C_e) = C'\big|_{\eta=0}, \tag{5.13a}$$

---

† Simpler, because the boundary evaluations occur on a *fixed* line $\bar{x} = 0$ of the $(\bar{x},\bar{t})$-plane.

$$C'' + 2\left(\overline{\eta} + \dot{\xi}\sqrt{\overline{t}/D}\right)C' = 0, \tag{5.13b}$$

where primes denote ordinary derivatives with respect to $\overline{\eta}$. These equations (5.13) would be ordinary differential expressions if the common factor $\dot{\xi}(\overline{t}/D)^{1/2}$ were independent of time. Therefore, in the same spirit as with the general similarity equation (5.3c), assume that this factor is a constant λ. Integrating that relation, we recover Eq. (4.19) for the interface schedule, and Eq. (5.13b) becomes simply

$$C'' + 2(\overline{\eta} + \lambda)C' = 0. \tag{5.13c}$$

This differential equation is easily integrated twice to yield the general solution

$$C(\overline{\eta}) = B + A\,\mathrm{erfc}(\overline{\eta}+\lambda), \tag{5.14}$$

which depends on two arbitrary constants. These, as usual, are determined by auxiliary conditions, about which a short digression is warranted.

We return to the moving $(\overline{x},\overline{t})$-frame in which the initial and boundary conditions are

$$C(\overline{x},0) = C(\infty,\overline{t}) = C_{\infty} \quad \text{and} \quad C(0,\overline{t}) = C_e. \tag{5.15a}$$

These *three* conditions, displayed again in Fig. 5.1, collapse into the *two*

$$C(\overline{\eta} \to \infty) = C_{\infty} \quad \text{and} \quad C(\overline{\eta} \to 0) = C_e \tag{5.15b}$$

when expressed in terms of the similarity variable (5.12a).† To solve this mini-mystery of the disappearing boundary condition, we first note that all our previous examples on unbounded regions were such that *the*

---

† Regardless of how that may ever happen, it is clear that we have solved our problem because these two boundary conditions (5.15b) are sufficient to determine the constants A and B of the general solution (5.14). In fact, we easily get $B = C_{\infty}$, and A is again given by Eq. (4.20a). Further, insertion of this solution into the Stefan condition (5.13a) yields precisely the same Eq. (4.20b) for the growth constant λ. The reader should verify that the concentration distribution (5.14) has an inflection point at $\overline{\eta} = -\lambda$. Why?

initial condition coincided with the far-field condition, as is evident from Figs. 4.3, 4.4, and 4.6. This is precisely the effect of a similarity variable such as (5.12a): $\bar{\eta}$ goes to infinity when either $\bar{x} \to \infty$ (if $\bar{t} \neq 0$) or $\bar{t} \to 0$ (if $\bar{x} \neq 0$), and the error function's behavior (5.14) must be the same in either case. Consequently, it should be intuitively evident that the origin of coordinates plays a peculiar role. In fact, the existence of a driving force (or supersaturation, because $C_\infty \neq C_e$) is closely related to the discontinuous behavior of the conditions (5.15a) at the origin. This leads to the question of singularities of the diffusion equation.

For simplicity, we consider a one-phase diffusion problem with no moving boundaries. The cases of several phases [*e.g.*, the temperature-redistribution problem (4.12-16)] and the previous Stefan problem can be discussed similarly. Assume that arbitrary initial and (Dirichlet) boundary conditions are prescribed

$$C(x,0) = \varphi(x) \quad \text{and} \quad C(0,t) = C_0(t), \tag{5.16a}$$

as shown on the left-hand side of Fig. 5.2, and observe that the origin of coordinates is a point of discontinuity because, in general,

$$\lim_{x \to 0+} \varphi(x) \neq \lim_{t \to 0+} C_0(t). \tag{5.16b}$$

Introduce, then, the following change of coordinates

$$\eta = x/2\sqrt{Dt} \quad \text{and} \quad \tau = \sqrt{t}, \tag{5.17}$$

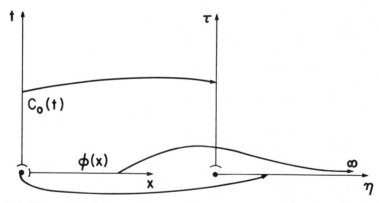

Fig 5.2: Illustration of the coordinate transformation (5.17) and the corresponding regions over which the auxiliary conditions must hold.

which somewhat generalizes our earlier considerations of similarity (*cf.* [5, Vol. I, pp. 330-338]. The jacobian of this transformation is easily computed: $\partial(\eta,\tau)/\partial(x,t) = 1/(4t\sqrt{D})$, and it is thus singular when $t = 0$. Therefore, viewed as a mapping of the (x,t)-plane onto the (η,τ)-plane, we expect "one-to-oneness" to break down along the x-axis. Specifically, inspection of the transformation (5.17) leads to the following conclusions: (i) The whole positive x-axis (origin excluded) collapses onto the point at infinity of the η-axis; (ii) the t-axis (origin excluded) maps onto the τ-axis (origin excluded); (iii) the origin of the (x,t)-plane maps onto the whole η-axis (origin included, point at infinity excluded). [The inquisitive reader should investigate what happens to other features of the (x,t)-plane, such as the the negative x-axis, the lines $t = $ const, and the phase boundary's schedule (4.19).] Consequently, the discontinuity (5.16b) is "smeared out" over the whole η-axis (which is helpful), but all the information on the initial condition $\varphi(x)$ must be contained in the single point at infinity, and this can happen *only if φ reduces to a constant.*

Next, we ask what form the diffusion equation takes in the new variables (5.17). A simple calculation, similar to that implied by Eqs. (5.12b), leads to

$$2\tau\frac{\partial C}{\partial \tau} = \frac{\partial^2 C}{\partial \eta^2} + 2\eta\frac{\partial C}{\partial \eta}, \tag{5.18}$$

and we still have a linear equation, albeit with variable coefficients. The main point, however, is that it reduces, when $\tau = 0$, to an ordinary differential equation

$$\frac{d^2 C}{d\eta^2} + 2\eta\frac{dC}{d\eta}, \tag{5.19a}$$

analogous to Eq. (5.13c). Considered as a two-point boundary-value problem on the positive η-axis, its solution is clearly

$$C(\eta) = \varphi(0+) + [C_0(0+) - \varphi(0+)]\,\mathrm{erfc}(\eta). \tag{5.19b}$$

This solution, identical to that of most of our previous examples, also shows its validity, for all time, if the initial condition φ and the boundary condition C_0 are independent of x and t, respectively. In other words, the first two conditions (5.15a) merge, and the solution (5.19b) "translates," unchanged, along the τ-axis if the initial and boundary conditions are constant.

To conclude this section, it should be emphasized that such an error-function solution (5.19b) is inadequate to describe other situations. For example, we recall that Eq. (4.21d) described a physical circumstance for which $C_0 \propto \sqrt{t}$. Such analytic expressions, however, are often used to "start" finite-difference schemes [9-12], even on bounded regions. These, we hinted in Sects. 1.4-5, often experience difficulties for the first few time steps, precisely because of discontinuities of the type (5.16b). Finally, we ought to ask if the variables (5.17) are the only possible choice. A partial answer will be given in Sects. 5.4-5, where we address some questions about the nonlinear diffusion equation.

Exercise 5.1: *Similarity in higher dimensions*
Express the rotationally symmetric diffusion equation (1.36) in terms of the similarity variable $\eta = r/2\sqrt{Dt}$. Find its general solutions. ∎

5.3 TWO APPLICATIONS OF SIMILARITY TO SPHERES

Our first application will be to the *impurity diffusion in a sphere*. It is the analog, for a bounded region, of the impurity diffusion problem (4.8). Consider a sphere of radius R that contains a mobile species at an initial constant level C_∞. Place that sphere in a furnace, say, so that diffusion occurs at noticeable rate, and assume that the ambient causes the impurity's surface concentration to take a constant equilibrium value C_e. Thus, we must solve the diffusion equation in the *interior* of the sphere with the initial and boundary conditions

$$C(r,0) = C_\infty, \quad 0 < r < R, \tag{5.20a}$$

$$C(R,t) = C_e, \quad t > 0, \tag{5.20b}$$

and we are left wondering what to do at the sphere's center, $r = 0$, where the diffusion equation (1.36) is singular.† In such cases, physical arguments suggest that C ought to be finite at the center.

---

† Until now we have never worried about this possibility, because all previous computations regarding spheres involved only the sphere's *exterior*.

The solution is easily found because, from Exercise 1.10, we know that the function $u = r(C + \text{const})$ is a solution of the 1-d diffusion equation. Accordingly, define a new unknown function u such that

$$C(r,t) = C_e + (C_\infty - C_e)\frac{u(r,t)}{r},$$ (5.21a)

which satisfies the initial and boundary conditions

$$u(r,0) = r, \quad 0 < r < R,$$ (5.21b)

$$u(R,t) = 0, \quad t > 0,$$ (5.21c)

$$\lim_{r\to 0} u/r < \infty.$$ (5.21d)

Because diffusion occurs in a bounded region $[0,R]$,† we look to the general similarity solution (5.7) for guidance. The sphere's radius is a natural length scale L_0, and we write this solution in trigonometric form

$$u \propto e^{-\lambda^2 Dt/R^2}[A\sin(\lambda r/R) + B\cos(\lambda r/R)].$$ (5.22a)

The coefficient B must vanish, or else we could not satisfy the condition (5.21d), and the other boundary condition (5.21c) requires that

$$\sin\lambda = 0, \quad \text{or} \quad \lambda_n = n\pi,$$ (5.22b)

where n is a positive integer.‡ The general solution to our problem is thus a superposition of harmonics (5.22a), and we write the Fourier series

$$u(r,t) = \sum_{n=1}^{\infty} A_n e^{-n^2\pi^2 Dt/R^2}\sin(n\pi r/R),$$ (5.22c)

---

† The following calculations are typical of problems where variables separate. The method goes back to Fourier's monumental work [13]. He discovered that trigonometric series can be used to represent diffusion phenomena, as well as periodic solutions of the wave equation.

‡ The case $n = 0$ is excluded because Eq. (5.21d) would be indeterminate; negative integers n are unnecessary, because they do not label linearly independent "eigenfunctions."

in which it remains only to determine the "Fourier amplitudes" A_n. This is easily done with the initial condition (5.21b), which now reads

$$r = \sum_{n=1}^{\infty} A_n \sin(n\pi r/R). \tag{5.23a}$$

Multiplying this equation by $\sin(m\pi r/R)$, integrating over the interval $[-R,R]$ and using the known "orthogonality"

$$\int_{-R}^{R} \sin(n\pi r/R) \sin(m\pi r/R)\, dr = R\delta_{n,m} \tag{5.23b}$$

of the trigonometric functions, we find that

$$A_n = -\frac{2R(-1)^n}{n\pi}, \tag{5.23c}$$

and thus the complete solution

$$C(r,t) = C_e + 2(C_\infty - C_e)$$

$$\times \sum_{n=1}^{\infty} \frac{(-1)^{n+1} e^{-\left(n\pi\sqrt{Dt}/R\right)^2} \sin(n\pi r/R)}{(n\pi r/R)}. \tag{5.24}$$

It is instructive to compute the rate of impurity outdiffusion

$$J_0 = -D\frac{\partial C}{\partial r}\Big|_{r=R} = 2D(C_\infty - C_e)R^{-1} \sum_{n=1}^{\infty} e^{-\left(n\pi\sqrt{Dt}/R\right)^2}, \tag{5.25a}$$

where we have differentiated the concentration distribution (5.24) with respect to r. This function is displayed (full line) in Fig. 5.3, and it should be compared with the following approximate calculation (dotted line) for short times. The latter allows a comparison with the earlier calculation (4.8c) for a half space (dashed line). If the diffusion length $\sqrt{Dt}$ is small compared to the sphere's radius R, then the impurity cannot distinguish a sphere from any other shape: All bodies look alike in the "tangent plane." In that case, the exponent of Eq. (5.25a) varies slowly with n, and, without any inquiry into rigor, we can replace the sum over n by an integral over the continuous variable $\xi = n\pi\sqrt{Dt}/R$. Thus, we get

Fig. 5.3: Flux from the surface of a sphere as a function of time. The dotted line corresponds to the approximation (5.25b), and the dashed line to Eq. (4.8c).

$$J_0 \cong 2D(C_\infty - C_e)\frac{1}{\pi\sqrt{Dt}}\int_{\pi\sqrt{Dt}/R}^{\infty} e^{-\xi^2}d\xi$$

$$= (C_\infty - C_e)\sqrt{D/\pi t}\,\mathrm{erfc}(\pi\sqrt{Dt}/R), \qquad (5.25b)$$

where the second equality results simply from the definition (4.10) of the complementary error function. For small argument of the error function, this is nothing but Eq. (4.8c) with a change in sign, since the r-axis points out of the sphere. Approximations of this type for exponential sums are very useful numerically when their exponents vary slowly.

Exercise 5.2: *Short-time solution*
Verify that the distribution (5.24) for the spherical problem reduces to Eq. (4.8b) for the planar problem when the diffusion length is small. [Hint: Use the second integral derived in Exercise 4.3.] ∎

The *Frank-Zener theory of precipitate growth* [14,15] provides a second application of similarity solutions to spheres. This theory provides the kinetics of growth for isolated, rotationally symmetric phases. Here, we restrict our attention to spheres. Recall that the approximate treatment of this problem in Sect. 2.5 assumed the existence of a quasi-steady-state solution, but that the surface condition was allowed to vary in accordance with local equilibrium, including capillarity effects. The present theory develops an exact time-dependent solution, but the boundary conditions are drastically simplified. Specifically, this theory neglects the surface tension γ. Thus, we must solve the same Eqs. (2.20-21) as before, except that the surface concentration C_R is assumed to be truly a constant C_e.

This is also a Stefan problem, and, taking a cue from the analogous planar case in Sects. 4.3 and 5.2, we write a trial solution

$$C(r,t) = C_\infty + A\,\frac{u(r,t)}{r}, \qquad r > R(t), \tag{5.26a}$$

where, as was observed previously, u must be a solution of the 1-d diffusion equation. We must now find that concentration distribution in the unbounded region *exterior* to the sphere. The similarity solutions (5.9), with r substituted for x, are now obvious candidates. But what value of n will do?† Since everything must be expressed in terms of the similarity variable $\eta = r/2\sqrt{Dt}$, it follows, by inspection of Eq. (5.26a), that $n = 1$, and the concentration distribution must be of the form

$$C(r,t) = C_\infty + A\,\frac{i^1\mathrm{erfc}(\eta)}{\eta}, \qquad \eta > R(t)/2\sqrt{Dt}. \tag{5.26b}$$

We have, again, two unknowns: the constant A above, and the surface's schedule $R(t)$. These are determined through the boundary conditions (2.21b,c)

$$C\,|_{r=R(t)} = C_e \quad \text{and} \quad \dot{R}(C_s - C_e) = D\frac{\partial C}{\partial r}\,\Big|_{r=R(t)}, \tag{5.26c}$$

repeated here for convenience. As in Sect. 4.3, the only way to satisfy both of these conditions is again to set

$$R(t) = 2\lambda\sqrt{Dt}, \tag{5.27a}$$

---

† The letter n, now used to label error functions, must be carefully distinguished from its previous use as a label for eigenfunctions.

i.e., a parabolic schedule. Inserting this schedule into Eqs. (5.26c), using the similarity form (5.26b), the formula (4.27), and the recursion (4.30b), we find the two unknowns

$$A = -\lambda \frac{C_\infty - C_e}{\mathrm{i}^1 \mathrm{erfc}(\lambda)}, \tag{5.27b}$$

$$2\sqrt{\pi}\,\lambda^2 e^{\lambda^2}\, \mathrm{i}^1 \mathrm{erfc}(\lambda) = \sigma, \tag{5.27c}$$

where σ is again the absolute supersaturation. This problem is formally solved because, given a driving force σ, we solve the transcendental equation (5.27c) for the growth constant λ; hence we get the schedule (5.27a), the constant A from Eq. (5.27b), and then the full concentration distribution (5.26b).†

At this point we must step back and ask what we have gained over the simple solution presented in Sect. 2.5. We have gained an exact solution to a moving-boundary problem that generalizes our previous estimate. To see this, assume that the initial radius R_0 in Eq. (2.25b) is very small. That estimate for the schedule is then $R = (2\sigma Dt)^{1/2}$. But this is precisely the limit of Eq. (5.27a) when the supersaturation is small. Indeed, using Eq. (4.32), we find from Eq. (5.27c) that $2\lambda^2 \cong \sigma$ for small supersaturations. Further, for each value of σ in the (semi-open) interval $[0,1)$ we can solve Eq. (5.27c) numerically for λ. Figure 5.4 shows how the function $\lambda(\sigma)$ deviates from its small-σ values as $\sigma \to 1$. The restriction on possible values of σ is explored in the next

Exercise 5.3: *Analytic results for the Frank-Zener theory*
Using the expansion (4.32), show that a quasi-steady state is obtained for small σ. In particular, examine the concentration distribution. Next, with the expansion (4.33a) and Exercise 4.8, show that there are no real solutions for $\sigma \geq 1$ and $\sigma < 0$. [It is recommended to sketch the function on the left-hand side of Eq. (5.27c).] Interpretation? ∎

---

† The original papers by Frank and Zener derive these results by solving the diffusion equation written in the similarity variable η, much as we had done in Eqs. (5.12-13). In fact, these authors analyze rotationally symmetric cases in any number d of dimensions. [*cf.* Exercise 5.1 above.]

Fig. 5.4: For the Frank-Zener problem, the deviation of the growth constant from its expression for small supersaturations.

No matter what the initial conditions might be, the Frank-Zener calculation certainly represents an asymptotic solution for spherical precipitate growth as $t \to \infty$. On the other hand, we have lost several physical features of the precipitation problem. The most glaring are the similarity solution's inability to handle capillarity (and thus critical nucleation considerations) and its related requirement of vanishing initial radius. Thus, in this problem, a sphere can only grow. Analyses of the effects of finite initial size, capillarity, and surface kinetics are mostly numerical. Reference [16] is one of many more recent studies of this topic. Finally, as with our earlier treatment, it is implicitly assumed that growth is shape preserving, *e.g.*, spheres become spheres, and we have again evaded the issue of morphological stability during this diffusion process.

5.4 BOLTZMANN'S TRANSFORMATION AND VARIABLE DIFFUSIVITY

A world in which all transport coefficients were constant would be very poor indeed. In connection with the random-walk problem that was presented in the very beginning, we had emphasized that any variation of the transition frequencies along sites would lead to a variable diffusivity. In fact, Exercise 1.3 provided a simple model for such an occurrence, and large external force fields are bound to affect D, as is clear from Eqs. (3.18). Another way to variable D's is through nonequilibrium thermodynamics [17], which takes the view that phenomenological relations, such as Fick's first law (1.12) and (1.31), must relate flux to gradients of chemical potential, rather than to concentration gradients. After all, constancy of μ, rather than of C, is required at equilibrium. Thus, we write

$$\mathbf{J} = -L\nabla\mu, \tag{5.28a}$$

which defines the Onsager coefficient L [*cf.* Eq. (1.48)]. If, further, the chemical potential of a solution can be expressed as in Eq. (2.17b), namely,

$$\mu = \mu^0(p,T) + kT\ln(\gamma C), \tag{5.28b}$$

where γ is the concentration-dependent activity coefficient,† then a comparison of Eqs. (5.28) and (1.31) shows that

$$\mathbf{J} = -D\nabla C, \quad \text{with} \quad D = \frac{LkT}{C}\left(1 + \frac{\partial\ln\gamma}{\partial\ln C}\right). \tag{5.28c}$$

Therefore, no matter what the concentration dependence of the Onsager coefficient L might be, the diffusivity D is certainly variable, as Darken first showed [18], because the activity coefficient is generally a function of the fields p, T, and C.‡

The world of variable diffusivity is, however, considerably more complicated to handle analytically, mainly because the diffusion equation (1.15)

$$\frac{\partial C}{\partial t} = \frac{\partial}{\partial x}\left(D\frac{\partial C}{\partial x}\right), \tag{5.29}$$

---

† Traditionally, the activity a is represented by the product γC, which defines the activity coefficient. Collisions in notation are unavoidable. Thus, γ represents either this coefficient or the surface tension, just as μ stands for either the chemical potential or the mobility coefficient, and L for either an Onsager coefficient or a length scale.

‡ There can be other reasons for variable diffusivity: In inhomogeneous media D can depend explicitly on position, and the action of external driving forces, such as temperature ramps (*cf.* Sect. 4.5), can cause D to depend on time. This last case will be considered further in Chap. 7.

repeated here for convenience, is nonlinear. In essence, we lose the ability to superpose elementary solutions, either as sums or as integrals. Consequently, distinct methods are expected for each given $D(C)$ dependence and for every set of initial and boundary conditions. In other words, at most, we can hope to find *particular* solutions.

Let us first examine *nonlinear diffusion in a half space under an arbitrary Dirichlet condition*

$$C(0,t) = C_0(t), \tag{5.30a}$$

together with an arbitrary initial condition

$$C(x,0) = \varphi(x) \tag{5.30b}$$

that matches a (constant) far-field condition

$$C(\infty,t) = C_\infty = \lim_{x \to \infty} \varphi(x). \tag{5.30c}$$

This is the nonlinear analog of the problem investigated in Sect. 5.2. We now ask whether Boltzmann's transformation (5.12a) is at all applicable. Consider, then, a mild generalization of the transformation (5.17)

$$\tau = At^\alpha \quad \text{and} \quad \eta = x/(Bt^\beta), \tag{5.31a}$$

where α and β are *arbitrary* exponents, and A and B are constants that can be conveniently chosen.† Calculating the partial derivatives as in Sect. 5.2, we transform Eq. (5.29) into

$$B^2\alpha\tau\frac{\partial C}{\partial \tau} = t^{1-2\beta}\frac{\partial}{\partial \eta}\left(D\frac{\partial C}{\partial \eta}\right) + B^2\beta\eta\frac{\partial C}{\partial \eta}. \tag{5.31b}$$

For the right-hand side to depend exclusively on η we require that the exponent of t vanish. Therefore, $\beta = \frac{1}{2}$, as with Boltzmann's transformation, but α (≥ 0) and A remain arbitrary because the left-hand side is homogeneous in τ. Choosing $B = 2$, we get

$$4\alpha\tau\frac{\partial C}{\partial \tau} = \frac{\partial}{\partial \eta}\left(D\frac{\partial C}{\partial \eta}\right) + 2\eta\frac{\partial C}{\partial \eta}, \quad \text{with} \quad \eta = x/2\sqrt{t}, \tag{5.32}$$

---

† The exponent α must be non-negative, or else τ would blow up as $t \to 0$.

i.e., an equation very similar to Eq. (5.18), which was studied earlier. Moreover, the transformation (5.31a) has the same singularities and effect on the auxiliary conditions as does the transformation (5.17). Therefore, if the surface concentration and the initial values are constant, and only under these circumstances, then the system of equations (5.32) and (5.30) reduces to the solution of the ordinary (but nonlinear), second-order differential equation for $C(\eta)$

$$\frac{d}{d\eta}\left(D\frac{dC}{d\eta}\right) + 2\eta\frac{dC}{d\eta} = 0, \tag{5.33a}$$

under the conditions

$$C(0) = C_0 \quad \text{and} \quad C(\infty) = C_\infty. \tag{5.33b}$$

There exist very few closed-form solutions of these equations (5.33). The search consists essentially in finding the necessary functional $D(C)$ relation compatible with a possible solution $C(\eta)$ [19,20]. On the other hand, Eqs. (5.33) are the basis of a very powerful numerical procedure, the *Boltzmann-Matano scheme* [21], for finding the diffusivity's concentration dependence from experimental data. This, in fact, was Boltzmann's goal.

Interdiffusion, phase transformations, and reactions at phase boundaries are facts of life for the metallurgist. These questions are reviewed in Refs. [22] and [23] for metallic systems and silicides, respectively. In addition, device engineers are sometimes concerned with *heterostructures,* a fancy name for a sequence of compositionally distinct, thin, semiconductor films. In that case, one is concerned that interdiffusion not alter these layers during subsequent processing steps [24,25]. To model these problems may require numerical methods because the phase boundaries are often neither stationary nor at local equilibrium. Furthermore, significant diffusion occurs over large portions of thin films. On the other hand, analytic modeling, when successful [26-28], is mainly based on Boltzmann's transformation, which necessarily leads to equations similar to Eqs. (5.33).

5.5 ANALYTIC SOLUTIONS FOR VARIABLE DIFFUSIVITY

The search for analytic solutions of Eq. (5.29) requires further special-ization. For example, there are many physical cases (*e.g.*, diffusion in porous media or in heavily doped semiconductors) where D varies as some power ν of C.† Consequently, let us consider *power-law diffusion in a half space*, such that no diffusant is present initially and far from the surface. Thus,

$$C(x,0) = C(\infty,t) = 0 \quad \text{for positive } x \text{ and } t, \tag{5.34a}$$

but we do *not* yet specify the boundary condition at the surface $x = 0$. The diffusivity now has the form

$$D(C) = D_*(C/C_*)^\nu, \tag{5.35}$$

where C_* and D_* are some corresponding reference values. Here, we follow the treatment of Pattle [29], as expanded by Boyer [30],‡ and we seek similarity solutions in the sense of Eqs. (5.1). To simplify their algebra, we will guess the form (power laws in t) of the scales $L(t)$ and $T(t)$; thus we assume that

$$C(x,t) = At^\alpha F(\eta), \quad \text{with} \quad \eta = x/(Bt^\beta), \tag{5.36}$$

where, again, the exponents α and β must be found, and the constants A and B will be chosen conveniently.

A few remarks concerning the exponents α, β, and ν are in order. First, it should be clear that α is quite arbitrary, but that β must be strictly positive, or else the presumptive form (5.36) could not satisfy the conditions (5.34a). As we have seen in Sect. 5.2, these two condi-tions collapse into the single condition

$$F(\infty) = 0 \tag{5.34b}$$

on the unknown function $F(\eta)$. Next, the case of constant diffusivity corresponds to the exponent $\nu = 0$ in Eq. (5.35). Further, negative ν's

---

† Recall Exercise 2.2; ν need not be an integer.

‡ The prehistory of this type of problem fades into World War II. It appears that solutions were first published by Zel'dovich and Kompaneets in the USSR (see Ref. [31]). Nevertheless, since these problems have bearing on detonations, it is likely that the solutions to follow were known to those skilled in the art of making bombs.

are physically unreasonable, since they would imply an infinite diffusivity at zero concentration. Finally, cases of positive ν correspond to simultaneous vanishing of D and C. Thus, we expect a steep diffusion front when the concentration is low. From a mathematical point of view, vanishingly small diffusivities in equations of the type (5.29) mean that the term containing the highest derivative vanishes. Such circumstances, in the theory of differential equations, are called *singular*, and they are closely related to the existence of such fronts.

We proceed, as before, by computing the partial derivatives

$$\partial/\partial x = (Bt^\beta)^{-1} d/d\eta \quad \text{and} \quad \partial C/\partial t = At^{\alpha-1}(\alpha F - \beta\eta F').$$

Inserting these, together with Eq. (5.35), into Eq. (5.29), we get

$$(D_* A^\nu / C_*^\nu B^2) t^{\alpha\nu+1-2\beta} (F^\nu F')' + \beta\eta F' - \alpha F = 0.$$

Let us now choose the leading coefficient so that this equation becomes as simple as possible an ordinary differential equation. Choosing

$$(D_*/C_*^\nu)(A^\nu/B^2) = 1, \tag{5.37a}$$

and satisfying the condition

$$\alpha\nu + 1 - 2\beta = 0, \tag{5.37b}$$

we get the ordinary differential equation

$$(F^\nu F')' + \beta\eta F' - \alpha F = 0 \tag{5.38}$$

for the unknown function $F(\eta)$. Given a physical problem (*i.e.*, for given ν), the conditions (5.37) relate the constant A to B and the exponent α to β. The as yet unspecified boundary condition will provide another set of conditions. We now consider three different cases, only the last of which yields a solution in closed form.

Case 1: Constant Dirichlet condition
Here, we impose a constant concentration

$$C|_{x=0} = C_0 = \text{const} \tag{5.39a}$$

at the free boundary. It follows from the form (5.36) that

$$At^\alpha F(0) = C_0. \tag{5.39b}$$

Consequently, the exponent α must vanish for this equation to be always true. With the condition (5.37b), this implies

$$\alpha = 0, \quad \beta = \frac{1}{2}, \tag{5.40a}$$

and we are still free to choose A and B. For example, if we take $A = C_0$, then Eq. (5.39b) is a boundary condition

$$F(0) = 1 \tag{5.41}$$

on the unknown function F, and Eq. (3.34b) provides another. Furthermore, since a unit of concentration, C_0, is available, we can choose this to be the reference concentration C_*, and we denote by D_0 the corresponding diffusivity. Their insertion into the condition (5.37a) determines these constants:

$$A = C_0, \quad B = \sqrt{D_0}. \tag{5.40b}$$

Therefore, the differential equation (5.38) reduces to

$$(F'F')' + \frac{1}{2}\eta F' = 0, \quad \text{with} \quad \eta = x/\sqrt{D_0 t}, \tag{5.42}$$

which is akin to Eq. (5.33a), discussed in connection with the Boltzmann transformation. Unfortunately, Eq. (5.42) is no more integrable in closed form than is Eq. (5.33a), but it is important to remember that we have reduced the diffusion equation to an ordinary differential equation in a similarity variable. This can be easily solved numerically under the two boundary conditions (5.34b) and (5.41). Readers familiar with hydrodynamics [1,2] will recognize an analogy (which runs deep) with the analysis of boundary layers.

Even if no analytic solution is available, valuable information can be gained directly from what we know about the exponents. For example, the surface flux $J_0 = -DC_x|_{x=0}$ has a time dependence which we determine from Eqs. (5.36) and (5.40),

$$J_0(t) = -F'(0)C_0\sqrt{D_0/t}, \tag{5.43}$$

and this expression is almost identical with Eq. (4.8c) for constant diffusivity.

Case 2: Constant Neumann condition
This second case concerns constant flux conditions at the free surface. It is very similar to the first case, and it also does not provide an integrable differential equation. This is the object of the next

Exercise 5.4: *Case of constant Neumann condition*
Assume that the surface flux is a given constant J_0. Determine the exponents (α, β) and choose the constants (A, B) appropriately. How does the surface concentration vary with time? ∎

Evidently, the cases of variable, but given, surface concentration $C_0(t)$ or surface flux $J_0(t)$ are handled in analogous manner, as long as these dependences are power laws in time.

Case 3: Instantaneous point source
This elementary case corresponds to the finite-difference calculation of Sect. 1.5 and to its continuum analog, the gaussian (4.3): A unit source is initially placed at the surface and is then allowed to relax by diffusion into the bulk. With the form (5.36), the total mass

$$M = \int_0^\infty C(x,t)\,dx = At^\alpha Bt^\beta \int_0^\infty F(\eta)\,d\eta \tag{5.44}$$

is conserved. Consequently, for M to be a constant M_0, independent of time, we require that $\alpha = -\beta$, and we can always choose $AB = M_0$. Using the previous conditions (5.37), we get

$$A = \left[\frac{C_*^\nu M_0^2}{D_*}\right]^{1/(\nu+2)} \quad, \qquad B = \left[\frac{D_* M_0^\nu}{C_*^\nu}\right]^{1/(\nu+2)} \quad, \tag{5.45a}$$

$$\alpha = -\beta = -1/(\nu + 2). \tag{5.45b}$$

Inserting these values into the differential equation (5.38), we can integrate once [because $\eta F' + F = (\eta F)'$] to obtain

$$(\nu + 2)\, F^\nu F' + \eta F = \text{const.}$$

Now we need the result of the next

Exercise 5.5: *A boundary condition for nonlinear diffusion*
If mass is conserved, show, using Eqs. (5.29) and (5.44), that the flux
at the origin must vanish. Is this a general result, or is it only true for
power-law diffusivities? ∎

The integration constant vanishes because of this exercise, and the
differential equation is integrable once again because it is separable:

$$F^\nu = \frac{1}{2}\frac{\nu}{\nu+2}(\eta_0^2 - \eta^2),$$

where η_0 is another integration constant. Next, the function F must be
real, so that, taking the νth root, we get

$$F(\eta) = \begin{cases} [\nu/2(\nu+2)]^{1/\nu}(\eta_0^2 - \eta^2)^{1/\nu} & \text{if } \eta \leq \eta_0, \\ 0 & \text{if } \eta > \eta_0. \end{cases} \tag{5.46}$$

Finally, to determine η_0 we insert this function (5.46) into the mass
conservation condition (5.44) to get

$$1 = \int_0^\infty F(\eta)\, d\eta = [\nu/2(\nu+2)]^{1/\nu} \int_0^{\eta_0} (\eta_0^2 - \eta^2)^{1/\nu}\, d\eta. \tag{5.47a}$$

The integral in question is a special case of the so-called beta function,
itself expressible in terms of gamma functions. This will be clear from
Eqs. (6.37) of the next chapter, because, putting $\xi = \eta^2/\eta_0^2$, the
integral is simply

$$\int_0^{\eta_0} (\eta_0^2 - \eta^2)^{1/\nu}\, d\eta = \frac{1}{2}\eta_0^{(\nu+2)/\nu} \int_0^1 (1-\xi)^{1/\nu} \xi^{-1/2}\, d\xi$$

$$= \frac{1}{2}\eta_0^{(\nu+2)/\nu} B(1+1/\nu, 1/2)$$

$$= \frac{1}{2}\eta_0^{(\nu+2)/\nu} \frac{\Gamma(1+1/\nu)\,\Gamma(1/2)}{\Gamma(3/2+1/\nu)}. \tag{5.47b}$$

With the recursion (4.23a) for gamma functions, it follows from
Eqs. (5.47) that

$$\eta_0 = \left[\frac{2(\nu+2)^{\nu+1}}{\nu\pi^{\nu/2}}\right]^{1/(\nu+2)} \left[\frac{\Gamma(1/2+1/\nu)}{\Gamma(1/\nu)}\right]^{\nu/(\nu+2)}, \tag{5.48}$$

which depends only on ν. This case is entirely solved, because the function F follows from (5.46) if we know the integration constant η_0, which we do from Eq. (5.48). Consequently, we know the full concentration distribution (5.36) because Eqs. (5.45) determine all the coefficients.

Exercise 5.6: *Limiting cases for small and large exponents ν*
With Stirling's formula (4.24b), compute η_0 for small ν. Then, show that the concentration tends towards (twice) the gaussian (3.3). Why "twice"? Compute these limits also when $\nu \rightarrow \infty$. ∎

It is instructive to rewrite this solution (5.36), (5.46) in the form

$$C(x,t) = C_0(t) \left[1 - \left(\frac{x}{L(t)} \right)^2 \right] \quad \text{for } x \le L(t), \qquad (5.49)$$

with the understanding that the concentration vanishes otherwise. By virtue of Eqs. (5.45), the surface concentration varies with time according to

$$C_0(t) = At^\alpha F(0) = \left[\frac{\nu \eta_0^2}{2(\nu + 2)} \right]^{1/\nu} \left[\frac{C_*^\nu M_0^2}{D_* t} \right]^{1/(\nu+2)}, \qquad (5.50a)$$

and the length scale is

$$L(t) = Bt^\beta \eta_0 = \eta_0 \left[\frac{D_* t M_0^\nu}{C_*^\nu} \right]^{1/(\nu+2)}. \qquad (5.50b)$$

This function is plotted in Fig. 5.5 for increasing values of the diffusivity exponent ν, and several of its features deserve comment. First, it is clear that the concentration vanishes when $x > L(t)$. In other words, there exists a moving front, akin to a propagating wave, that separates disturbed regions from others which have not yet felt the influence of the initial point source. This is in marked contrast with the gaussian (4.3), its analog for cases of constant diffusivity, for which *all* regions of space are immediately affected (albeit infinitesimally, it is true, for distant points) by an initial point disturbance. Further, when the exponent ν is large, the concentration distribution is quite flat over large portions of its domain. When ν is small, Fig. 5.5 indicates that we

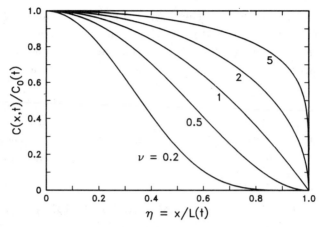

Fig. 5.5: Scaled concentration distribution for power-law diffusivity. Note the abrupt front that develops when ν is large.

approach a gaussian shape, as should be the case. This assertion is proved in Exercise 5.6 above. Next, the surface concentration (5.50a) decreases like $t^{-1/(\nu+2)}$, and its dependence on the total mass M_0 is quite different: linear in the gaussian case, and to the $[2/(\nu+2)]$th power when $\nu \neq 0$. The same differences appear for the length scale: independent of M_0 in the gaussian case, and to the $[\nu/(\nu+2)]$th power otherwise. Again, both C_0 and $L(t)$ converge to gaussian behavior when $\nu = 0$. Indeed, the reader is urged to consult the excellent treatise by Zel'dovich and Raizer [31], particularly their Chap. 10, for a discussion of this problem by dimensional analysis.

In sum, we have shown that nonlinear diffusion presents features resembling traveling waves and that its characteristic parameters (concentration and length scales) may depend on the total mass in unexpected ways. Needless to say, there are very few other nonlinear problems that can be solved in closed form (Ref. [31] presents another: an initial "dipole" disturbance), and there is every reason to ask why our attention should not center exclusively on numerical solutions, rather than on these isolated analytical solutions. There is, however, an important asymptotic result in the case of linear diffusion: *Any* initial disturbance, in an infinite or semi-infinite medium, will evolve by diffusion into a gaussian if we should observe the concentration distribution in faraway regions. This somewhat awkward statement is given a precise meaning and is proved in Appendix C. The beauty of asymptotic methods, however, resides in their application to nonlinear problems, as well. In a remarkable paper [32], Zel'dovich and Barenblatt prove an analogous result for power-law diffusion: The solution we have found for a point source can also be regarded as the asymptotic behavior of an arbitrary initial disturbance. In other words, concentration "tails" for nonlinear diffusion can be fitted to a known analytic expression.

Exercise 5.7: *Moments and power-law diffusion*
Consider the moments (C.8) defined in Appendix C. Explore the consequences of their constancy for power-law diffusion in a half space. ∎

REFERENCES

1. H. Schlichting, *Boundary-Layer Theory* (McGraw-Hill, New York 1968).
2. G. Birkhoff, *Hydrodynamics: A study in logic, fact and similitude,* 2nd ed. (Princeton University Press, Princeton, NJ 1960).
3. N.N. Lebedev, *Special Functions and their Applications* (reprinted by Dover, New York 1972).
4. G.W. Bluman and J.D. Cole, *Similarity Methods for Differential Equations* (Springer-Verlag, Berlin 1974).
5. W.F. Ames, *Nonlinear Partial Differential Equations in Engineering,* Volumes I and II (Academic Press, New York 1965 and 1972).
6. L. Dresner, *Similarity Solutions of Nonlinear Partial Differential Equations* (Pitman Advanced Publishing Program, Boston 1983).
7. L. Boltzmann, "Zur Integration der Diffusionsgleichung bei variabeln Diffusionscoefficienten," *Ann. der Phys.* **53**, 959 (1894).
8. G. Rosen, "Galilean Invariance and the General Covariance of Nonrelativistic Laws," *Amer. J. Phys.* **40**, 683 (1972).
9. G.W. Evans II, E. Isaacson, and J.K.L. MacDonald, "Stefan-like Problems," *Quart. Appl. Math.* **8**, 312 (1950).
10. B. Boley, "A General Starting Solution for Melting and Solidifying Slabs," *Int. J. Engng. Sci.* **6**, 89 (1968).
11. M.B. Small and R. Ghez, "Growth and Dissolution Kinetics of III-V Heterostructures formed by LPE," *J. Appl. Phys.* **50**, 5322 (1979).
12. R. Ghez, "Expansions in Time for the Solution of One-Dimensional Stefan Problems of Crystal Growth," *Int. J. Heat Mass Transfer* **23**, 425 (1980).

13. J. Fourier, *Théorie Analytique de la Chaleur* (1822) [see the annotated edition of Fourier's early manuscript by I. Grattan-Guinness, *Joseph Fourier 1768-1830*, MIT Press, Cambridge, MA 1972].

14. F.C. Frank, "Radially Symmetric Phase Growth controlled by Diffusion," *Proc. Roy. Soc.* **201A**, 586 (1950).

15. C. Zener, "Theory of Growth of Spherical Precipitates from Solid Solution," *J. Appl. Phys.* **20**, 950 (1949).

16. R.J. Schaeffer and M.E. Glicksman, "Fully Time-Dependent Theory for the Growth of Spherical Crystal Nuclei," *J. Cryst. Growth* **5**, 44 (1969).

17. S.R. de Groot and P. Mazur, *Non-equilibrium Thermodynamics* (North-Holland, Amsterdam 1962).

18. L.S. Darken, "Diffusion, Mobility and their Interrelation through Free Energy in Binary Metallic Systems," *Trans. AIME* **175**, 184 (1948).

19. J.R. Philip, "General Method of Exact Solution of the Concentration-Dependent diffusion equation," *Australian J. Phys.* **13**, 1 (1960); see also *ibid.*, p. 13.

20. B. Tuck, "Some Explicit Solutions to the Non-linear Diffusion Equation," *J. Phys.* **D9**, 1559 (1976).

21. B. Tuck, *Introduction to Diffusion in Semiconductors*, pp. 199-203 (Peter Peregrinus, Stevenage 1974).

22. M. Hillert, "Diffusion and Interface Control of Reactions in Alloys," *Metall. Trans.* **6A**, 5 (1975).

23. F.M. d'Heurle and P. Gas, "Kinetics of Formation of Silicides: A Review," *J. Mater. Res.* **1**, 205 (1986).

24. L.L. Chang and A. Koma, "Interdiffusion between GaAs and AlAs," *Appl. Phys. Lett.* **29**, 138 (1976).

25. K.S. Seo, P.K. Bhattacharya, G.P. Kothiyal, and S. Hong, "Interdiffusion and Wavelength Modification in $In_{0.53}Ga_{0.47}As/In_{0.52}Al_{0.48}As$ Quantum Wells by Lamp Annealing," *Appl. Phys. Lett.* **49**, 966 (1986).

26. G.V. Kidson, "Some Aspects of the Growth of Diffusion Layers in Binary Systems," *J. Nucl. Mater.* **3**, 21 (1961).

27. C. Wagner, "The Evaluation of Data obtained with Diffusion Couples of Binary Single-Phase and Multiphase Systems," *Acta Met.* **17**, 99 (1969).

28. D.S. Williams, R.A. Rapp, and J.P. Hirth, "Phase Suppression in the Transient Stages of Interdiffusion in Thin Films," *Thin Solid Films* **142**, 65 (1986).

29. R.E. Pattle, "Diffusion from an Instantaneous Point Source with a Concentration-Dependent Coefficient," *Quart. J. Mech. and Appl. Math.* **7**, 407 (1959).

30. R.H. Boyer, "On some Solutions of a Non-linear Diffusion Equation," *J. Math. and Phys.* **40**, 41 (1961).

31. Ya.B. Zel'dovich and Yu.P. Raizer, *Physics of Shock Waves and High-Temperature Hydrodynamic Phenomena*, Vol. II (Academic Press, New York 1967).

32. Ya.B. Zel'dovich and G.I. Barenblatt, "The Asymptotic Properties of Self-Modelling Solutions of the Nonstationary Gas Filtration Equations," *Dokl. Akad. Nauk SSSR* **118**, 4 (1958) [Engl. transl. in *Sov. Phys. Doklady* **3**, 44 (1958)].

Chapter 6

A USER'S GUIDE TO THE LAPLACE TRANSFORM

This chapter provides a compact, self-contained introduction to the Laplace transform, a powerful technique that can be applied with ease to a wide variety of time-dependent problems. Rather than strive for the greatest generality, this introduction addresses only those simplest properties that are most useful for the solution of diffusion problems. Only elementary methods of calculus are used, and integration in the complex plane is avoided altogether. It follows that questions regarding analyticity, the inversion integral, and asymptotic behavior cannot be completely answered, although some such properties will be demonstrated.

6.1 MOTIVATION, DEFINITION, AND SOME EXAMPLES

Rational operations, supplemented by raising to powers, do not exhaust all possible operations on functions. The basic operations of differentiation and integration are cornerstones of analysis. They are sometimes called transcendental operations because they are defined through limiting processes. It would be convenient to translate these operators into a form that resembles multiplication and division. Then, for example, repeated differentiation might correspond to repeated multiplication in the new language. The Laplace transform [1-3], one of many useful *integral transforms* [2,4], has precisely this property. Its systematic use, in the context of the diffusion equation, allows us to *reduce by one* the number of independent variables and, thus, to reduce a given problem to a far simpler one. We begin with a

Definition: Let $f(t)$ be a function (real or complex) of the real variable t. Its Laplace transform, denoted $\mathscr{L}\{f(t)\}$ or $\overline{f}(p)$, is defined by the integral

$$\mathscr{L}\{f(t)\} = \overline{f}(p) = \int_0^\infty e^{-pt} f(t)\, dt. \tag{6.1}$$

The integral (6.1) clearly transforms functions of t (time, say) into new functions of a new variable p, the Laplace variable. That is to say, for each value of p for which the above integral converges, we get the corresponding value of the transformed function.

For example, the Laplace transform of an arbitrary exponential $f(t) = e^{\alpha t}$ is

$$\mathscr{L}\{e^{\alpha t}\} = \int_0^\infty e^{-(p-\alpha)t}\, dt = \frac{1}{p - \alpha}, \tag{6.2}$$

and the exponential function has been transformed into the new function $\overline{f}(p) = (p-\alpha)^{-1}$. This example, elementary though it may seem, can teach us quite a lot about the Laplace transform. First, we notice that the transform of a transcendental function of t (the exponential) is a rational function (thus much simpler) in the Laplace variable p. Second, we carelessly performed the integration (6.2) as if this were always possible. But that integral will not converge if $p \leq \alpha$ (for real p and α), and the Laplace transform then cannot exist. In other words, the definition (6.1) requires still a condition on the values of p (real or complex) that ensures convergence of the integral. For this first example, the condition on the real parts is $\mathscr{R}p > \mathscr{R}\alpha$. Third, we have just learned that the image function $\overline{f}(p)$ is generally a complex function of the complex variable p and that the defining integral (6.1) converges in some right-hand plane of the complex p-plane. Fourth, the original function $f(t)$ can also be complex. For example, taking $\alpha = i\omega$ in the exponential and separating the real and imaginary parts, we get

$$\int_0^\infty e^{-(p-i\omega)t}\, dt = \int_0^\infty e^{-pt}(\cos \omega t + i \sin \omega t)\, dt$$

$$= \frac{1}{p - i\omega} = \frac{p}{p^2 + \omega^2} + i\frac{\omega}{p^2 + \omega^2}, \tag{6.3}$$

whence the transforms of the trigonometric functions, which are de-fined for $\mathcal{R}p > 0$. These transforms are recorded in Table 6.1 at the end of this chapter. Another special case of the exponential emerges when $\alpha = 0$. The function $f(t)$ is a constant everywhere equal to unity, and, from Eq. (6.2), we have $\mathcal{L}\{1\} = 1/p$.

Table 6.1, placed at the end of this chapter, is an illustration of the *mapping* provided by Eq. (6.1). To each function $f(t)$ for which that integral makes sense (*i.e.*, converges) there corresponds a single image $\bar{f}(p)$. But is that correspondence one-to-one? Evidently not, because values of $f(t)$ for *negative* t's are irrelevant to the definition (6.1). In other words, two functions that differ only on the negative t-axis have the same image. For example, our exponential can be anything at all to the left of the origin. Because the Laplace transform was designed to deal with processes that develop from some initial condition, one simply restricts (or defines) the class of original functions $f(t)$ to those that are identically zero for $t < 0$.† Heaviside's function

$$H(t) = \begin{cases} 0 & \text{if } t < 0, \\ 1 & \text{if } t > 0, \end{cases} \tag{6.4a}$$

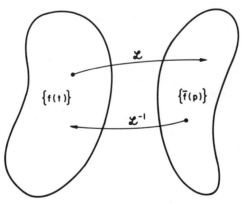

Fig. 6.1: Schematic representation of the Laplace transform and its inverse.

---

† There are other more subtle restrictions on the class of original functions that provide a one-to-one mapping. Two functions that differ by an amount whose integral (6.1) vanishes must be considered equal. Nevertheless, for functions f that are continuous on the positive real axis, except for countably many jump discontinuities, their Laplace transforms determine the f's uniquely, except for the values at the jumps.

whose Laplace transform

$$\mathscr{L}\{H(t)\} = 1/p \tag{6.4b}$$

is the same as the unit function's, evidently falls in that class. It can be used to "truncate" any function that does not vanish for negative t's. Thus,

$$\bar{f}(p) = \mathscr{L}\{f(t)H(t)\} \tag{6.5}$$

defines this one-to-one correspondence, regardless of f's behavior to the left of the origin.† Figure 6.1 portrays the mapping $\mathscr{L}$ prescribed by Eq. (6.1), as well as its inverse $\mathscr{L}^{-1}$ whenever this can be reckoned.

6.2 ELEMENTARY PROPERTIES AND FURTHER EXAMPLES

The Laplace transform enjoys many interesting and useful properties that we will exploit. Among these are linearity, change of scale, translations, the effect of differentiation and integration, convolution, and asymptotic behavior. These properties serve to derive new transform pairs from known ones. We shall survey only the most important ones for our purpose.

A. Linearity
This property is so evident from the definition (6.1) that we even failed to point it out in the derivation of Eqs. (6.3). By linearity of the integral operator we have

$$\mathscr{L}\{\alpha f(t) + \beta g(t)\} = \alpha\mathscr{L}\{f(t)\} + \beta\mathscr{L}\{g(t)\} \tag{6.6}$$

for any scalars‡ α and β and any functions $f(t)$ and $g(t)$ for which the indicated operations make sense. The inverse operator $\mathscr{L}^{-1}$ is likewise linear. For example, expanding our exponential $e^{\alpha t}$ in Taylor series

---

† The multiplicative function $H(t)$ in Eq. (6.5) is most often implied. This generally does not cause difficulties, except when one wishes to find transforms of functions with retarded arguments, in accordance with Eq. (6.10) below. [For instance, see the calculation that follows Eq. (6.24).]

‡ In these sections, real variables or parameters (most often positive) are denoted by Latin letters, reserving Greek letters for complex quantities. The only notable exception is the Laplace variable p which can be complex.

around the origin, integrating term by term (without concern for rigor), and expanding the right-hand side of Eq. (6.2) in a geometric series, we get

$$\mathscr{L}\{e^{\alpha t}\} = \int_0^\infty dt\, e^{-pt} \sum_{n=0}^\infty \frac{(\alpha t)^n}{n!} = \sum_{n=0}^\infty \frac{\alpha^n}{n!} \mathscr{L}\{t^n\}$$

$$= \frac{1}{p-\alpha} = p^{-1} \sum_{n=0}^\infty (\alpha/p)^n.$$

Identifying equal powers of α, we then obtain the transform of integer powers of t:

$$\mathscr{L}\{t^n\} = n!/p^{n+1}, \quad n \geq 0. \tag{6.7a}$$

Exercise 6.1: *Transform of arbitrary powers*

Show, directly from the definition (6.1) of the Laplace transform and from the definition (4.22) of the gamma function, that

$$\mathscr{L}\{t^\nu\} = \Gamma(\nu+1)/p^{\nu+1}, \quad \nu > -1. \tag{6.7b}$$

In particular, compute explicitly the cases $\nu = \pm\frac{1}{2}$. ∎

B. Change of scale

If we know the Laplace transform of a function $f(t)$, can we then infer the transform of $f(at)$? As the following simple calculation shows, the answer is affirmative if a is positive. By definition

$$\mathscr{L}\{f(at)\} = \int_0^\infty e^{-pt} f(at)\, dt = a^{-1} \int_0^\infty e^{-(p/a)s} f(s)\, ds$$

$$= \frac{1}{a}\bar{f}(p/a), \tag{6.8}$$

where the second equality results from the substitution $s = at$. The restriction on a follows because the dummy variables t and s must both run over the same half axis $[0,\infty]$.

The next properties are equally easy to verify. Thus, they are recorded without proof and left as Exercises 6.2-3 below.

C. Translations

The observant reader will have noticed that the transform of the exponential $e^{\alpha t}$ is obtained from that of Heaviside's function by a mere translation $p \to p - \alpha$ of the Laplace variable. This feature is quite general: If $\bar{f}(p)$ is the transform of $f(t)$, then

$$\mathcal{L}\{e^{\alpha t} f(t)\} = \bar{f}(p - \alpha). \tag{6.9}$$

In other words, multiplication by exponentials in the original space corresponds to translations of the independent variable p in image space. For example, from Eq. (6.7a) we immediately infer that $\mathcal{L}\{e^{\alpha t} t^n\} = n!/(p-\alpha)^{n+1}$.

There exists a second "shifting" theorem: Translations of the original variable t induce multiplication by exponentials in image space. Let $g(t) = f(t-a)$ be the "retarded" function $f(t)$ shifted a units to the right, with the understanding that $g \equiv 0$ for $t < a$. This situation is illustrated in Fig. 6.2. We then have

$$\mathcal{L}\{f(t-a)H(t-a)\} = e^{-ap}\bar{f}(p), \quad a \geq 0. \tag{6.10}$$

For example, the Laplace transform of $H(t-a)$, Heaviside's function with a unit step at $t = a$, is simply $p^{-1}e^{-ap}$. Equations (6.9) and (6.10) illustrate a general feature of Laplace transforms, namely, that theorems often occur in "dual" pairs.

Exercise 6.2: *Simple proofs*

Prove the Properties (6.9-10). What can you say about these statements if α or a is negative? Can α be complex, and if so, what is the

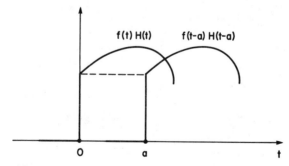

Fig 6.2: An arbitrary function, and the same function with retarded argument.

condition for convergence? [Hint: Assume that f is bounded.] ▮

We now come to the most useful feature of the Laplace transform: its effect on

D. Differentiation

The basic question is: Can we compute the transform of df/dt if we know the transform of $f(t)$? Under fairly general conditions, if $\dot{f}$ is substituted in the definition (6.1), then integration by parts yields

$$\mathscr{L}\{\dot{f}(t)\} = p\mathscr{L}\{f(t)\} - f(0+). \tag{6.11}$$

In other words, the transform of a derivative corresponds to a multiplication by p in image space, with the initial value thrown in as an added bonus. Note that the *value* of $f(t)$ at the origin is irrelevant; only its limit to the right enters into this formula. For example, take $f(t)$ to be Heaviside's function (6.4a). Its derivative vanishes for all positive t. The left-hand side of Eq. (6.11) then also vanishes, and, in view of the result (6.4b), the right-hand side also vanishes identically because $H(0+) = 1$. This example is telling because the value $H(0)$ is undefined.

The differentiation property (6.11) is essential for the application of the Laplace transform to the diffusion equation (1.14). If, now, we identify the variable t with time, and if we recognize that the definition (6.1) does not affect the space variables, *e.g.*,

$$\mathscr{L}\{C_x\} = \partial(\mathscr{L}\{C\})/\partial x$$

— tersely, the $\mathscr{L}$-operator and x-derivatives commute — then the transformed concentration $\overline{C}(x,p) = \mathscr{L}\{C(x,t)\}$ obeys the *ordinary* differential equation

$$D\frac{d^2\overline{C}}{dx^2} - p\overline{C} = -C(x,0+), \tag{6.12}$$

which is surely much more easily solved than the original partial differential equation (1.14). These considerations apply regardless of the number of space dimensions and of the coordinates used to describe them. In sum, the Laplace transform reduces by one the number of independent variables in any linear differential expression, and the Laplace variable p is then just a parameter. The main task, however, is the computation of the inverse transform that takes the solution $\overline{C}(x,p)$ of Eq. (6.12) into the space of original functions $C(x,t)$.

Exercise 6.3: *More simple proofs*
Supply the proof of the property (6.11), and think of an example where it may not apply. [Hint: What happens if f has jump discontinuities on the half axis $t > 0$?] ∎

The basic operation (6.11) can be applied to higher derivatives. For example, repeated application of that rule to the function $g = \dot{f}$ gives

$$\mathscr{L}\{\ddot{f}(t)\} = \mathscr{L}\{\dot{g}\} = p\mathscr{L}\{\dot{f}(t)\} - \dot{f}(0+)$$

$$= p^2\mathscr{L}\{f(t)\} - pf(0+) - \dot{f}(0+). \tag{6.13}$$

For example, we derive De Moivre's formula $e^{it} = \cos t + i \sin t$, which was surreptitiously used in the calculations leading to Eq. (6.3). This will also illustrate how we can easily solve any linear differential equation with constant coefficients. Consider the function $f(t) = e^{it}$. We verify that it satisfies the differential equation $\ddot{f} + f = 0$ and the initial conditions $f(0) = 1$, $\dot{f}(0) = i$. Applying Eq. (6.13) and the linearity property (6.6) to the differential equation, we immediately get an *algebraic* expression for the transform $\bar{f}(p) = (p+i)/(p^2+1)$. Hence, reading "backwards" the first two entries of Table 6.1, we get the desired formula.

E. Integration
The operation inverse to differentiation is integration

$$g(t) = \int_0^t f(s)\,ds, \tag{6.14}$$

and one safely predict that, if differentiations on the original space of functions correspond essentially to multiplications by p on the image space, then integrations, such as Eq. (6.14), will correspond to divisions by p of their transforms. This result is an immediate consequence of the differentiation property (6.11) because $\dot{g} = f$ and $g(0+) = 0$. Consequently,

$$\mathscr{L}\{\dot{g}(t)\} = p\mathscr{L}\{g(t)\} - g(0+) = p\mathscr{L}\{g(t)\},$$

and the desired result follows:

$$\mathscr{L}\left\{\int_0^t f(s)\,ds\right\} = \frac{\bar{f}(p)}{p}. \qquad (6.15)$$

This formula is often used in reverse: We can evaluate the inverse transform of an expression such as $\bar{f}/p$ if we know the corresponding pair $f(t)$ and $\bar{f}(p)$. For example, the formula (6.7a) for powers of t follows by repeated integration of $H(t)$.

There also exist properties dual to (6.11) and (6.15), namely, multiplication and division by t on the space of original functions correspond to differentiation and integration with respect to p on the space of image functions [1-3]. We shall not dwell on these properties, because they are rarely used in the context of linear diffusion.

Before continuing to examine transforms that are of interest for diffusion, we pause briefly to consider the delta function $\delta(t)$ and its transform. There are many ways to interpret this function, mainly because it is not a function at all in the usual sense, but rather a construct. We illustrate this with our exponential. Consider the family of functions

$$g_a(t) = ae^{-at}H(t), \quad t \geq 0, \; a > 0, \qquad (6.16)$$

that depends continuously on the real positive parameter a. As functions of t, these functions have a maximum, a, that increases and a time constant, a^{-1}, that decreases with a. Nonetheless, the integral over all time of this family is a constant. Indeed, we easily verify that

$$\lim_{a \to \infty} g_a(t) = \begin{cases} 0 & \text{if } t > 0, \\ \infty & \text{if } t = 0, \end{cases} \qquad (6.17a)$$

and

$$\int_0^\infty g_a(t)\,dt = 1 \quad \forall a > 0. \qquad (6.17b)$$

In other words, this family exhibits a singularity at the origin $t = 0$ if a is large, and that singularity is integrable. This should sound familiar from our work with the gaussian (4.3), which, we recall, is singular in the space variable x, time t playing there the role of a parameter. Now the essential property of the delta function is its ability to select a particular functional value from a definite integral. This property is indeed shared by the family (6.16), for, integrating by parts over an arbitrary function f

$$\int_0^\infty g_a(t)f(t)\,dt = f(0+) + \int_0^\infty e^{-at}\dot{f}(t)\,dt,$$

and taking the limit of large a, we get

$$\lim_{a \to \infty} \int_0^\infty g_a(t)f(t)\,dt = f(0+). \tag{6.17c}$$

Consequently, there is some justification in setting

$$\lim_{a \to \infty} ae^{-at}H(t) = \delta(t). \tag{6.18a}$$

Further, since $\mathscr{L}\{g_a(t)\} = a/(p+a)$ follows from Eqs. (6.2) and (6.6), we then have

$$\mathscr{L}\{\delta(t)\} = \lim_{a \to \infty} a/(p + a) = 1. \tag{6.18b}$$

Thus, the transform of the delta function is *independent* of p. None of our other examples has this feature, and none ever will, because, again, the delta function is defined through a limiting process on a family of functions. The limit does not exist in the sense of a continuous function, as should be clear from Eq. (6.17a), but its effect on an integral is nonetheless well defined.†

Exercise 6.4: *Selection property of delta functions*
Show, with any suitable family of functions, that

---

† The existence of many distinct families of functions of a real parameter a, like $g_a(t)$, each of which has a constant transform in the limit when that parameter a increases without bound [*cf.* Exercise 6.5], drives another nail in the coffin of the appellation "delta *function*." Strictly speaking, with the selection property (6.17d) taken as an equivalence relation, $\delta(t)$ is the *equivalence class* of such limits.

$$\int_0^\infty f(t)\delta(t-a)\,dt = f(a) \tag{6.17d}$$

for any integrable function f and for any positive number a. ∎

Exercise 6.5: *More delta functions*
Consider the family of functions $g_\varepsilon(t) = 1/\varepsilon$ for $0 \le t < \varepsilon$, and which vanish elsewhere. Show that they have all the properties (6.17) in the limit $\varepsilon \to 0$. In what sense have we defined a derivative of $H(t)$? Can you generalize to other similar families that vanish outside an interval? ∎

6.3 LAPLACE TRANSFORMS IN $\sqrt{p}$

We now have the basic tools to evaluate a large number of inverse transforms that may arise in practice. For example, if $\bar{f}(p)$ is a rational function of p, then a partial-fraction expansion, together with the translation property (6.9) and Eqs. (6.7), allow its inversion. Likewise, arbitrary inverse powers of p, perhaps multiplied by exponentials in p, can be inverted using Exercise 6.1 and the translation property (6.10).

Of major concern to us, however, are solutions of the transformed diffusion equation (6.12). Its fundamental solutions are of the form $\exp[\pm x(p/D)^{1/2}]$, which are not yet in our lexicon of available transforms. We begin by stating and proving the following

Lemma: If we know the Laplace transform $\bar{f}(p)$ of the function $f(t)$, then the inverse transform of $\bar{f}(\sqrt{p})$ (*i.e.*, the *same* function of the square-root argument) is given by the integral

$$\mathscr{L}^{-1}\{\bar{f}(\sqrt{p})\} = \frac{1}{2\sqrt{\pi}\,t^{3/2}}\int_0^\infty s e^{-s^2/4t} f(s)\,ds. \tag{6.19}$$

We verify this lemma by first applying the $\mathscr{L}$-operator to both sides of Eq. (6.19). Then we interchange the s- and t-integrations, and perform the t-integration through the change of variables $t \to \xi = s/2\sqrt{t}$. Thus

$$\mathcal{L}\left\{\frac{1}{2\sqrt{\pi}\,t^{3/2}}\int_0^\infty s e^{-s^2/4t}f(s)\,ds\right\} = \int_0^\infty ds\,\frac{sf(s)}{2\sqrt{\pi}}\int_0^\infty \frac{dt}{t^{3/2}}e^{-pt}e^{-s^2/4t}$$

$$= \int_0^\infty ds\,\frac{2f(s)}{\sqrt{\pi}}\int_0^\infty d\xi\,e^{-ps^2/4\xi^2}e^{-\xi^2} = \int_0^\infty e^{-s\sqrt{p}}f(s)\,ds,$$

where the integral over ξ is exactly the third result of Exercise 4.3. We now apply this lemma to a number of transforms that we have already encountered:

a. From the result (6.18b) and the second translation property (6.10), we know that $\mathcal{L}\{\delta(t-a)\} = e^{-ap}$. It follows immediately from the lemma (6.19), applied to the function $f(t) = \delta(t-a)$, and the selection property (6.17c) that

$$\mathcal{L}^{-1}\{e^{-a\sqrt{p}}\} = \frac{a}{2\sqrt{\pi}\,t^{3/2}}e^{-a^2/4t}. \tag{6.20}$$

Likewise, from Eq. (6.4b) for Heaviside's function, we have $\mathcal{L}\{H(t-a)\} = p^{-1}e^{-ap}$. Application of the lemma, once again, gives

$$\mathcal{L}^{-1}\{p^{-1/2}e^{-a\sqrt{p}}\} = \frac{1}{2\sqrt{\pi}\,t^{3/2}}\int_a^\infty s e^{-s^2/4t}\,ds$$

$$= \frac{1}{\sqrt{\pi t}}e^{-a^2/4t}, \tag{6.21a}$$

because we recognize that the integrand is

$$s e^{-s^2/4t} = -2t\frac{d}{ds}e^{-s^2/4t}. \tag{6.21b}$$

b. It should be noted that Eq. (6.21a) also results from Eq. (6.20) on integrating both sides of the latter with respect to the parameter a in the range $[a,\infty]$. The integration evidently commutes with the $\mathcal{L}$-operator. We thus have a general procedure for the computation of $\mathcal{L}^{-1}\{p^{-n/2}e^{-a\sqrt{p}}\}$. For example, from Eq. (6.21a) we have

$$\mathcal{L}^{-1}\{p^{-1}e^{-a\sqrt{p}}\} = \frac{1}{\sqrt{\pi t}}\int_a^\infty e^{-b^2/4t}\,db = \operatorname{erfc}(a/2\sqrt{t}), \tag{6.22}$$

where the last step results from a simple change of variables and the definition (4.10) of the complementary error function. We should now sense that these inverse transforms are related to the class of iterated error functions. This is the object of the next

Exercise 6.6: *Transforms of the iterated error functions*
By induction, show that

$$\mathscr{L}\{(2\sqrt{t})^n i^n \text{erfc}(a/2\sqrt{t}) = p^{-(1+n/2)}e^{-a\sqrt{p}}, \qquad \forall n \geq -1. \quad \blacksquare \quad (6.23)$$

Note that Eqs. (6.21-22) above are special cases of this result (6.23). Therefore, we know how to invert $e^{-a\sqrt{p}}$ divided by arbitrary *integer powers* of $\sqrt{p}$.† There are cases, however, where one must learn to invert this exponential divided by *polynomials* in $\sqrt{p}$. The next example is of this type, and it requires a little more work.

c. We know from Eqs. (6.2), and (6.4-5) that

$$\mathscr{L}^{-1}\left\{\frac{1}{p(p+h)}\right\} = \mathscr{L}^{-1}\left\{\frac{1}{h}\left[\frac{1}{p}-\frac{1}{p+h}\right]\right\}$$

$$= \frac{1}{h}\left[1-e^{-ht}\right]H(t), \qquad (6.24)$$

where h is any constant. From the second translation property (6.10), it follows that

$$\mathscr{L}^{-1}\left\{\frac{e^{-ap}}{p(p+h)}\right\} = \frac{1}{h}\left[1 - e^{-h(t-a)}\right]H(t-a),$$

for any positive constant a. Applying the lemma to this transform pair, we have

$$\mathscr{L}^{-1}\left\{\frac{e^{-a\sqrt{p}}}{\sqrt{p}(\sqrt{p}+h)}\right\} = \frac{1}{2h\sqrt{\pi}\,t^{3/2}}\int_a^\infty s e^{-s^2/4t}\left[1-e^{-h(s-a)}\right]ds,$$

and, with Eq. (6.21b), an integration by parts yields

---

† Inspection of Eq. (6.20) shows that we could extend formula (6.23) to integers $n < -1$ if we knew how to define the corresponding iterated error functions for such values of n. This is achieved by repeated differentiation of Eq. (6.21a) with respect to a. We obtain the class of Hermite polynomials $H_{-n-1}(a/2\sqrt{t})$ multiplied by gaussians, of which Eq. (6.20) is a member.

$$\mathscr{L}^{-1}\left\{\frac{e^{-a\sqrt{p}}}{\sqrt{p}(\sqrt{p}+h)}\right\} = \frac{e^{ha}}{\sqrt{\pi t}}\int_a^\infty e^{-s^2/4t}e^{-hs}\,ds.$$

We then note that the exponents on the right-hand side can be written as a square. A first change of variables $\xi = s/2\sqrt{t}$ makes this quite obvious, and a second change $u = \xi+h\sqrt{t}$ brings the integral into the form (4.10) for the complementary error function. We get finally

$$\mathscr{L}^{-1}\left\{\frac{e^{-a\sqrt{p}}}{\sqrt{p}(\sqrt{p}+h)}\right\} = e^{ha+h^2t}\operatorname{erfc}\left(\frac{a}{2\sqrt{t}} + h\sqrt{t}\right), \qquad (6.25)$$

where the parameter a must be non-negative, but h can be anything, real or complex. This last result is very useful because many other transform pairs follow, either on setting the parameters a and h to special values, or on differentiating or integrating with respect to these parameters.

Exercise 6.7: Two useful transforms
Find the inverse transforms of

$$\frac{e^{-a\sqrt{p}}}{\sqrt{p}+h} \quad \text{and} \quad \frac{e^{-a\sqrt{p}}}{p(\sqrt{p}+h)}$$

by differentiation and integration of Eq. (6.25) with respect to a. ∎

6.4 A SAMPLE CALCULATION

We apply the Laplace transformation to a simple problem that was solved earlier by direct methods: the impurity diffusion problem (4.8) in a half space under constant surface conditions. This example illustrates how the systematic use of the Laplace transform reduces the search for solutions to mere algebraic juggling, rather than to detailed knowledge of special functions' behavior. For convenience, we state the mathematical problem once again: Solve the diffusion equation

$$C_t = DC_{xx}, \quad x > 0, \ t > 0 \qquad (6.26a)$$

under the initial condition

$$C(x,0+) = C_\infty \tag{6.26b}$$

and the boundary conditions

$$C(0,t) = C_0, \quad C(\infty,t) = C_\infty, \quad t > 0, \tag{6.26c,d}$$

where C_0 (called C_e in Chap. 4) and C_∞ are constants. Transforming the diffusion equation and using the initial condition (6.26b), we get the ordinary differential equation (6.12) with constant coefficients

$$D\overline{C}'' - p\overline{C} = -C_\infty, \tag{6.27}$$

and, because the boundary conditions are constant in time, they transform into

$$\overline{C}(0,p) = C_0/p, \quad \overline{C}(\infty,p) = C_\infty/p, \tag{6.28a,b}$$

by virtue of Eq. (6.4b) and the linearity property (6.6). The general solution of Eq. (6.27) is evidently

$$\overline{C}(x,p) = p^{-1}C_\infty + Ae^{-x\sqrt{p/D}} + Be^{x\sqrt{p/D}}, \tag{6.29a}$$

where A and B are arbitrary constants with respect to x. They are, however, functions of the Laplace variable p. Indeed, the second transformed boundary condition (6.28b) shows that solutions must be bounded at large distances from the surface. This implies that $B = 0$ because, in general, $\mathscr{R}\sqrt{p} > 0$. On the other hand, the first condition (6.28a) allows us to find A, and we get

$$\overline{C}(x,p) = p^{-1}C_\infty + p^{-1}(C_0-C_\infty)e^{-x\sqrt{p/D}}. \tag{6.29b}$$

Hence, from the transforms (6.4b) and (6.22), or reading "backwards" the appropriate entries (4 and 10) in Table 6.1, we get the desired solution

$$C(x,t) = C_\infty + (C_0-C_\infty)\,\text{erfc}(x/2\sqrt{Dt}). \tag{6.30}$$

It is identical to Eq. (4.8b) if we note the relation (4.10a) between erf and erfc. Note also that the initial and boundary conditions need not be constants. This is illustrated in the next

Exercise 6.8: *Variable initial condition*

Assume that the initial condition of the above problem is an arbitrary function of x, *i.e.*, Eq. (6.26b) is replaced by

$$C(x,0+) = \varphi(x), \quad \text{with} \lim_{x \to \infty} \varphi(x) = C_\infty.$$

Assume, also, that the boundary conditions (6.26c,d) remain unaltered. Find the general solution in terms of a Green's function. [Hint: Solve the differential equation by the method of "variation of constants."] Compare your result with Eq. (4.5). ∎

The above procedure is quite general, and it can be applied to a wide variety of problems. To summarize: Apply the Laplace transform to any linear diffusion equation in 1-d and obtain an ordinary linear differential equation; the initial condition is, felicitously, "built in." Solve this equation with the *transformed* boundary conditions. Finally, using a table of transforms and, perhaps, some of the properties in Sect. 6.2, find the solution in the space of original functions. But what should one do if the table does not contain the desired entries? We return to the general question of inversion in the last section of this chapter.

6.5 THE CONVOLUTION THEOREM

This property is next in importance to the Laplace transform's ability to translate differentiation and integration into rational operations in image space. To motivate convolutions, consider the differential equation of a damped linear oscillator. Its schedule $x(t)$ satisfies

$$m\ddot{x} + \mu\dot{x} + kx = f(t), \tag{6.31a}$$

under the initial conditions

$$x(0+) = x_0 \quad \text{and} \quad \dot{x}(0+) = v_0. \tag{6.31b}$$

Here, m, μ, and k are the particle's mass and its friction and restoring constants, respectively, and $f(t)$ is an arbitrary impressed force. The solution of this linear second-order differential equation (6.31a) is well known; we recall its structure: It consists of the sum of the general solution $x_h(t)$ of the *homogeneous* equation (*i.e.*, when $f \equiv 0$) and *any* *particular* solution $x_p(t)$ of the full inhomogeneous equation. The general solution, here, depends on two arbitrary multiplicative constants. But these solutions, while interesting, do not describe physical situations, unless the initial conditions (6.31b) can be met. These conditions can be distributed quite arbitrarily over x_h and x_p. We shall now see that it is best to attribute them to the solutions of the homogeneous equation. Indeed, if we take the transform of Eq. (6.31a) and use the differentiation properties (6.11) and (6.13), then we get

$$\bar{x}(p) = \frac{mx_0 p + mv_0 + \mu x_0}{mp^2 + \mu p + k} + \frac{\bar{f}(p)}{mp^2 + \mu p + k}. \qquad (6.32)$$

It is now evident that the first term of Eq. (6.32), *i.e.*, the transform $\bar{x}_h(p)$, has been entirely specified by the initial conditions and that the second term $\bar{x}_p(p)$ stems from the *homogeneous* conditions

$$x(0+) = \dot{x}(0+) = 0. \qquad (6.33a)$$

In addition, we recognize the characteristic polynomial of the differential equation (6.31a) in both denominators of Eq. (6.32). Now, $\bar{x}_h(p)$ is a ratio of two polynomials in p, and it thus has a partial-fraction expansion. Consequently, its inverse $x_h(t)$ is available from our lexicon (see Exercise 6.9 below). The particular solution is another matter, because we might not know how to invert $\bar{f}(p)$, for arbitrary f's, divided by a second-degree polynomial. Let us assume for a moment that $\bar{f} = 1$. In this case the inversion of the second term would be similar to that of the first term. But its numerator corresponds precisely to the fifth entry $f(t) = \delta(t)$ in Table 6.1. As in the development of Eqs. (1.29) and (4.5), call *this* particular solution the Green's function $G(t)$. Because any impressed force f can be decomposed according to

$$f(t) = \int_0^t f(s)\,\delta(t-s)\,ds, \qquad (6.33b)$$

and because the differential equation (6.31a) is linear, it follows that the particular solution x_p, satisfying the homogeneous initial conditions (6.33a), is given by the integral

$$x_p(t) = \int_0^t f(s)\,G(t-s)\,ds. \qquad (6.33c)$$

Exercise 6.9: *Explicit solution of Eqs. (6.31)*
Cast the problem in dimensionless form by choosing appropriate time and distance scales. Find x_h and G explicitly. ∎

Results such as Eq. (6.33c) are familiar to us. They indicate that the effect of an arbitrary forcing term is the superposition, with retarded arguments, of elemental effects due to instantaneous point sources. This situation is so common that the right-hand side of Eq. (6.33c) is graced with a name: the Laplace *convolution* of the functions f and G, and it is denoted $f \star G$. More generally, one defines the convolution of two functions, f and g, as the integral

$$f \star g = \int_0^t f(s)\,g(t-s)\,ds. \qquad (6.34)$$

It represents a "product" operation on the space of integrable functions, as shown in the next

Exercise 6.10: *Algebraic properties of convolutions*
Show that convolutions are commutative, associative, and distributive with respect to the addition of functions. (Such a set of operations goes a long way towards defining what is called an *algebra*.) ∎

Now what does all this have to do with Laplace transforms? The problem (6.31), after all, could have been solved by direct methods. Nevertheless, the development that led from Eq. (6.32) to (6.33c) shows that there must be a connection between convolutions in the original space of functions and products in image space. Indeed, we have the so-called *convolution theorem*

$$\mathscr{L}^{-1}\{\bar{f}(p)\,\bar{g}(p)\} = f \star g, \qquad (6.35)$$

which we now verify. Take the $\mathscr{L}$-operation of both sides and evaluate

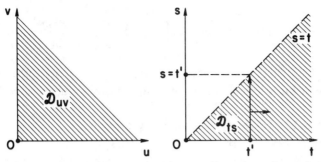

Fig. 6.3: Corresponding domains under the coordinate transformation required for the proof of the convolution theorem.

the left-hand side:

$$\bar{f}(p)\,\bar{g}(p) = \int_{\mathscr{D}_{uv}} e^{-p(u+v)} f(u)\, g(v)\, du\, dv = \int_{\mathscr{D}_{ts}} e^{-pt} f(s)\, g(t-s)\, ds\, dt$$

$$= \int_0^\infty dt\, e^{-pt} \int_0^t ds\, f(s)\, g(t-s) = \mathscr{L}\{f \star g\}.$$

The first integral is nothing but the product of the two Laplace transforms, written as a double integral over the dummy variables (u,v). The domain $\mathscr{D}_{uv}$ of these variables extends over the first quadrant, as shown in Fig. 6.3. The change of variables $t = u + v$ and $s = u$ transforms this domain into the octant $\mathscr{D}_{ts}$. It is easily verified that the jacobian of the transformation is unity, whence the second integral. Finally, this octant is swept out by first integrating over s (from $s = 0$ to $s = t$), and then varying t over its whole range to yield the last integral.

This theorem has an immediate application to our sample problem of the last section. If, for example, the first boundary condition (6.26c) is an *arbitrary* function of time $C_0(t)$, then Eq. (6.29a) still holds, but the "constant" A will now be expressed in terms of the transform $\overline{C}_0(p)$. Thus, Eq. (6.29b) now reads

$$\overline{C}(x,p) = p^{-1}C_\infty + [\overline{C}_0(p) - C_\infty/p]\,e^{-x\sqrt{p/D}},$$

and, using the result (6.20), the convolution theorem (6.35), and the definition (4.10a), we get the desired solution

$$C(x,t) = C_\infty \, \text{erf}(x/2\sqrt{Dt}) + \int_0^t C_0(s) \frac{xe^{-x^2/4D(t-s)}}{2\sqrt{\pi D}\,(t-s)^{3/2}} \, ds. \quad (6.36)$$

Note that nowhere was it necessary to actually compute the transform $\overline{C}_0$ of the arbitrarily given concentration at the boundary $x = 0$.

Exercise 6.11: *To verify the boundary condition*
Can you merely put $x = 0$ into Eq. (6.36) to verify the boundary condition $\lim_{x \to 0} C(x,t) = C_0(t)$? [Hint: Find a good change of variables for the integral.] ∎

The convolution theorem has many other applications. For example, it can be used to solve certain *integral equations*, as will be illustrated in the last chapter. Here, we will only mention its application to the evaluation of some definite integrals. The first of these defines the so-called beta function

$$B(\mu,\nu) = \int_0^1 (1-s)^{\mu-1} s^{\nu-1} \, ds, \quad (6.37a)$$

where μ and ν are any real exponents greater than zero. This integral would be a convolution of the form (6.34) if the independent variable t were set to unity. The trick, therefore, is to imbed it in the larger class of integrals of the form

$$f(t) = \int_0^t (t-s)^{\mu-1} s^{\nu-1} \, ds, \quad (6.37b)$$

whose Laplace transform is

$$\overline{f}(p) = \mathcal{L}\{t^{\mu-1}\} \mathcal{L}\{t^{\nu-1}\} = \frac{\Gamma(\mu)\Gamma(\nu)}{p^{\mu+\nu}}$$

according to the convolution theorem (6.35) and the seventh entry of Table 6.1. The inverse transform follows from this same entry, and we get

$$f(t) = \frac{\Gamma(\mu)\Gamma(\nu)}{\Gamma(\mu+\nu)} \, t^{\mu+\nu-1}. \quad (6.37c)$$

Hence, setting $t = 1$, we have evaluated the beta function

$$B(\mu,\nu) = \Gamma(\mu)\Gamma(\nu)/\Gamma(\mu+\nu) \tag{6.37d}$$

entirely in terms of the gamma function.

Exercise 6.12: *The area of a circle*
Use the beta function to calculate the area of a circle. [Hint: Write the area as the integral $dx\,dy$ over the unit circle in the first quadrant. Then change variables to $\xi = x^2$ and $\eta = y^2$.] Can you generalize for the volume of a d-dimensional sphere? ∎

Another useful integral is that obtained by repeated "nested" integration (n-fold) of a given function

$$g_n(t) = \int_0^t dt_n \int_0^{t_n} dt_{n-1} \cdots\cdots \int_0^{t_2} dt_1\, f(t_1), \tag{6.38a}$$

where $t > t_n > t_{n-1} > \ldots > t_1 > 0$. We wish to express it as a single integral. The previous discussion on transforms of integrals provides a clue because Eq. (6.14) is of this form. Further, it can be written symbolically $g_1 = f \star 1$, and its Laplace transform (6.15) is indeed given by the convolution theorem (6.35). Now, the multiple integral (6.38a) can be written as a recursion

$$g_n(t) = \int_0^t g_{n-1}(t_n)\, dt_n = g_{n-1} \star 1, \tag{6.38b}$$

and its transform is

$$\bar{g}_n(p) = \frac{1}{p}\bar{g}_{n-1}(p) = \frac{1}{p^2}\bar{g}_{n-2}(p) = \ldots = \frac{1}{p^n}\bar{f}(p), \tag{6.38c}$$

by repeated application of the integration property (6.15), and because $g_0 = f$. Inverting this last with the convolution theorem and the sixth entry of Table 6.1, we get the desired result

$$g_n(t) = \int_0^t \frac{(t-s)^{n-1}}{(n-1)!} f(s)\, ds. \tag{6.38d}$$

We see once again that the Laplace transform translates transcendental operations (6.38a,b) into simple divisions (6.38c) on the space of image functions.

6.6 A FEW WORDS ON ASYMPTOTICS

It may happen that we are only interested in the behavior of a diffusive system, either for short times or in the limit of long times. On the other hand, many problems are intractable analytically except in these limits. Consequently, we may ask if there exist general methods to extract these asymptotic limits from a given problem's Laplace transform. The answer is affirmative, and the vast subject of asymptotics is related to the analytic properties of the Laplace integral (6.1) in the complex p-plane. For example, we have seen that this integral converges only in certain regions of that plane. In addition, the problem (6.31) of rectilinear motion shows that the complete solution (*cf.* Exercise 6.9) depends crucially on the partial-fraction expansion of the inverse of a second degree polynomial in p, *i.e.*, on the *poles* of that function. Therefore, a full discussion of asymptotic behavior requires a good dose of complex variable theory. Here, in keeping with our elementary discussion of diffusion, we content ourselves with a few remarks.

The first thing we ought to notice from the definition (6.1) of a Laplace transform is that $\bar{f}(p)$ vanishes when p is large: The original function $f(t)$ gets damped by the exponential for large positive values of p.† Indeed, all the tabulated transforms in this chapter vanish in this limit, except, of course, the transform of the delta function. But does that large-p limit correspond to anything of interest in the space of original functions? The sixth entry of Table 6.1 gives a clue: Convergence to zero of t^n corresponds to rapid decay for large p of its transform $n!p^{-(n+1)}$. Therefore, we anticipate that the behavior of $f(t)$ on the t-axis should be, in some sense, "reciprocal" to the behavior of $\bar{f}(p)$ in the p-plane. This statement can be substantiated if the original function has a convergent Taylor expansion around $t = 0+$. Then, as in the calculations leading to Eq. (6.7a), we get

---

† More generally, legitimate Laplace transforms must vanish at infinity along any ray in the right-hand p-plane.

$$\bar{f}(p) = \int_0^\infty dt\, e^{-pt} \sum_{n=0}^\infty \frac{t^n}{n!} f^{(n)}(0+) = \sum_{n=0}^\infty \frac{f^{(n)}(0+)}{n!} \mathscr{L}\{t^n\}$$

$$= \sum_{n=0}^\infty \frac{f^{(n)}(0+)}{p^{n+1}}, \tag{6.39}$$

with the usual notation $f^{(n)}$ for the nth derivative. In other words, the way in which $f(t)$ approaches its initial value is related to the way in which its transform $\bar{f}(p)$ behaves around the point at infinity. In particular, extracting the zeroth term from the sum, multiplying throughout by p, and taking the limit of large p, we get the so-called *initial-value theorem*

$$\lim_{p \to \infty} p\bar{f}(p) = \lim_{t \to 0+} f(t). \tag{6.40}$$

Exercise 6.13: *Initial- and final-value theorems*
Use the differentiation property (6.11) to prove, directly, the initial-value theorem (6.40). Likewise, prove the *final-value theorem*

$$\lim_{p \to 0+} p\bar{f}(p) = \lim_{t \to \infty} f(t). \quad \blacksquare \tag{6.41}$$

These results call for several remarks. In the first place, although the theorems are correct, we must remember that the above manipulations do not constitute rigorous proofs, since we have freely interchanged limits. Next, the initial- and final-value theorems (6.40-41) must be understood as follows: *If* the limits $f(0+)$ or $f(\infty)$ of the original functions are otherwise known to exist, *then* we can compute their values from limits of Laplace transforms. To proceed in the opposite direction can sometimes yield absurd results. For example, applying the final-value theorem (6.41) to the first entry of Table 6.1 leads to a perfectly reasonable left-hand limit, $\lim_{p \to 0} p^2/(p^2+\omega^2) = 0$, but the right-hand side, $\lim_{t \to \infty} \cos \omega t$, does not exist. Last, the expansion (6.39) can be understood as asymptotic rather than convergent. Indeed, integrating the definition (6.1) by parts, we get

$$\bar{f}(p) = \int_0^\infty e^{-pt} f(t)\, dt = -\frac{1}{p} e^{-pt} f(t)\, |_{t=0}^\infty + \frac{1}{p} \int_0^\infty e^{-pt} \dot{f}(t)\, dt.$$

This process can be repeated to yield a series of the form (6.39), except that now we have not assumed Taylor convergence of f.

The previous considerations were all questions of the following type: Given the behavior of the original function around some value of t, can we infer the behavior of the image function around some "reciprocal" value of p? These considerations can be elaborated. For example, one can show ([5, pp. 180-183] that the seventh entry of Table 6.1 holds asymptotically, namely, if

$$f(t) \sim e^{\alpha t} t^\nu \qquad \text{for } t \to 0+ \text{ or } t \to \infty, \tag{6.42a}$$

then

$$\bar{f}(p) \sim \Gamma(\nu+1)/(p-\alpha)^{\nu+1} \qquad \text{for } p \to \infty \text{ or } p \to \alpha+. \tag{6.42b}$$

In addition, the power-series expansion (6.39) admits an important generalization known as Watson's lemma [6, p. 34]: If the original function has an *asymptotic* expansion in arbitrary powers t^ν ($\nu > -1$), then its image function results by termwise transformation of these powers. Even this lemma can be generalized to asymptotic expansions in terms of "gauge functions" other than powers [7].

The converse question, however, is more interesting: Given the image function, or at least, its asymptotic behavior, can we recover the original function of time? That question is related to the problem of inversion of Laplace transforms [1-3], and, as was mentioned earlier, it cannot be addressed properly in the real domain. Here, the following remarks will suffice. First, if the transform $\bar{f}(p)$ admits a convergent series expansion in inverse powers (not necessarily integer) of p, then the original function can be recovered by formal termwise inversion (see Refs. [8], Chap. 8, especially p. 174, and [9], Theorem 2). Therefore, by construction, large-p expansions lead to small-t expressions of the original function. This procedure is sometimes applicable to convergent expansions in powers of functions other than $p^{-\nu}$ [3, pp. 198-200]. Second, the theorem (6.42) can often be used in reverse, namely, (a) follows from (b), when α is a singularity of the transform [3, pp. 234-238]. The precise conditions under which this is valid cannot be stated without a thorough examination of the properties of the "inversion integral." The interested reader will find more details in Refs. [8] (Chap. 7, especially pp. 150-154) and [9] (Theorem 3). Nevertheless, we will use or verify the previous statements in the examples to follow in the next chapter.

Table 6.1: Short table of Laplace transforms†

Original function $f(t)$‡	Image function $\bar{f}(p)$	Conditions §
1. $\cos \omega t$	$p/(p^2+\omega^2)$	
2. $\sin \omega t$	$\omega/(p^2+\omega^2)$	
3. $e^{\alpha t}$	$1/(p-\alpha)$	$\mathcal{R}p > \mathcal{R}\alpha$
4. 1	$1/p$	
5. $\delta(t)$	1	
6. t^n	$n!/p^{n+1}$	$n \geq 0$
7. $e^{\alpha t}t^\nu$	$\Gamma(\nu+1)/(p-\alpha)^{\nu+1}$	$\mathcal{R}p > \mathcal{R}\alpha,\ \nu > -1$
8. $\dfrac{a}{2\sqrt{\pi}\,t^{3/2}}e^{-a^2/4t}$	$e^{-a\sqrt{p}}$	$a > 0$
9. $\dfrac{1}{\sqrt{\pi t}}e^{-a^2/4t}$	$p^{-1/2}e^{-a\sqrt{p}}$	$a \geq 0$
10. $\operatorname{erfc}(a/2\sqrt{t})$	$p^{-1}e^{-a\sqrt{p}}$	$a \geq 0$
11. $(2\sqrt{t})^n \operatorname{i}^n\operatorname{erfc}(a/2\sqrt{t})$	$p^{-(1+n/2)}e^{-a\sqrt{p}}$	$a \geq 0,\ n \geq -1$
12. $e^{ha+h^2t}\operatorname{erfc}\left(\dfrac{a}{2\sqrt{t}} + h\sqrt{t}\right)$	$\dfrac{e^{-a\sqrt{p}}}{\sqrt{p}(\sqrt{p}+h)}$	$a \geq 0,\ \forall h$

† A very complete compilation is given in Ref. [10]. In addition, almost every book on Laplace transforms contains an adequate table, *e.g.*, [1-3].

‡ It is understood that all these functions vanish identically for negative t's in the spirit of Eq. (6.5).

§ Existence conditions for transforms, rational in p, are $\mathcal{R}p > 0$, unless otherwise stated. Only the negative real axis is excluded for transforms in $\sqrt{p}$: the phase of p can anything in $(-\pi,\pi)$.

REFERENCES

1. M.R. Spiegel, *Theory and Problems of Laplace Transforms* (Schaum Publishing Co., New York 1965).
2. R.V. Churchill, *Operational Mathematics*, 3rd ed. (McGraw-Hill, New York 1972).
3. G. Doetsch, *Introduction to the Theory and Application of the Laplace Transform* (Springer-Verlag, Berlin 1974).
4. C.J. Tranter, *Integral Transforms in Mathematical Physics*, 3rd ed. (Methuen & Co., Ltd, London 1966).
5. D.V. Widder, *The Laplace Transform* (Princeton University Press, Princeton, NJ 1941).
6. A. Erdélyi, *Asymptotic Expansions* (reprinted by Dover, New York 1956).
7. A. Erdélyi, "General Asymptotic Expansions of Laplace Integrals," *Arch. Rational Mech. Anal.* 7, 1 (1961).
8. G. Doetsch, *Handbuch der Laplace-Transformation*, Band II (Birkhäuser-Verlag, Basel 1972).
9. R.A. Handelsman and J.S. Lew, "Asymptotic Expansion of Laplace Convolutions for Large Argument and Tail Densities for certain Sums of Random Variables," *SIAM J. Math. Anal.* 5, 425 (1974).
10. A. Erdélyi, W. Magnus, F. Oberhettinger, and F. Tricomi, *Tables of Integral Transforms*, Vol. 1 (McGraw-Hill, New York 1954).

Chapter 7

FURTHER TIME-DEPENDENT EXAMPLES

The constructive (or synthetic) methods of Chaps. 4 and 5 often require considerable ingenuity. Solving a given diffusion problem depends on our ability to match certain known elementary solutions of the diffusion equation to the initial and boundary conditions. It thus requires detailed knowledge of these solutions' behavior. In contrast, this last chapter describes deductive (or analytic) methods, based mainly on the Laplace transform, that can be applied with ease to a wide variety of problems.

7.1 LASER PROCESSING

Sources of heat, such as lasers and electron beams, are commonly used to induce either structural modifications [1,2] or phase changes in engineering [3,4] and biological [5,6] materials. The many investigations of "laser annealing," especially for ion-implanted materials, have been excellently reviewed in Refs. [7] and [8]. In this chapter, we use a simplified form of this process to investigate means of computing interactions between external sources of heat and a solid material whose properties we wish to modify. Assume that we have such a material on which impinges a beam produced, say, by a laser. Assume, also, that the solid is semi-infinite and that it is bounded by a planar surface. Let $F_0(t)$ be the laser flux† incident on that surface; its time-dependence is, for the moment, arbitrary. Not all the impinging power is effective; if R is the material's reflection coefficient, then a fraction

---

† In this context, the word "flux" means the same thing as "areal power density"; both are measured in units of power per unit area. Here, we are also assuming that the beam has an infinite lateral extent.

RF_0 is reflected by the solid. The remainder, $(1-R)F_0$, transits into the solid where it dissipates. The interaction of matter and light is a very complex process; it can be described phenomenologically, however, by introducing the notion of *skin depth* λ. According to this simple picture, the change in dissipated power P (per unit volume) is proportional to the power itself, the coefficient of proportionality being precisely $1/\lambda$. Thus, $dP/dx = -\lambda^{-1}P$ leads to

$$P(x) = P_0 e^{-x/\lambda}, \tag{7.1}$$

where P_0 is the power dissipated at the solid's surface $x = 0$. In the absence of diffusion processes, the total power dissipated in the solid is the integral over all x of expression (7.1). This must balance exactly the net power input $(1-R)F_0$, and it follows that $P_0 = (1-R)F_0/\lambda$.

It is easy to write the appropriate equation for temperature changes because we recognize that Eq. (7.1) is a source term (in the sense of Sect. 1.8) of the diffusion equation that follows from Eqs. (1.37). Thus, for the case of constant coefficients, we must solve

$$\rho c \frac{\partial T}{\partial t} = K \frac{\partial^2 T}{\partial x^2} + (1-R)F_0(t)\lambda^{-1}e^{-x/\lambda}. \tag{7.2}$$

We assume that the solid is initially at a constant temperature T_∞ and that this temperature is somehow maintained far from the surface. Furthermore, we assume that the surface temperature is small enough so that radiation into the surrounding medium is negligible compared to the laser flux.† Therefore, we have the initial and boundary conditions

$$T(x,0) = T_\infty, \tag{7.3a}$$

$$T(\infty,t) = T_\infty, \qquad \frac{\partial T}{\partial x}\Big|_{x=0} = 0. \tag{7.3b,c}$$

The attentive reader may have noted the similarity between this problem and the heating problem (4.21). Here, the laser energy gives rise to a *distributed source* (7.1) that penetrates a distance λ into the solid,

---

† According to the Stefan-Boltzmann σT^4 radiation law, where the constant σ has the value 5.64×10^{-12} W/cm² K⁴, an upper bound to the emitted flux from a silicon surface at its melting point (1683 K) is only 45 W/cm². This should be compared to the power input, 0.1 J/cm² delivered in a nanosecond, of a typical pulsed laser.

and the surface flux (7.3c) vanishes. On the other hand, energy input was simulated earlier through a *surface source* (4.21a), expressed as a boundary condition.† Are these distinct representations of the same physical process? One would hope not, and a clue emerges if we recognize that the power dissipation (7.1) is proportional to $\lambda^{-1}\exp(-x/\lambda)$. This, according to Eqs. (6.16-18), is nothing but $\delta(x)$ in the limit of small λ's, *i.e.*, a unit surface source.

Exercise 7.1: *Heat balance and boundary conditions*
Compute an irradiated solid's change in total internal energy, and balance it against the power input from external sources. Use the heat equation (7.2), but retain the form (7.1) for the source. Discuss the boundary conditions when the skin depth is finite and when it vanishes. ∎

This problem is easily solved by Laplace transformation. First, it is best to recast it in a dimensionless form that eliminates many of the physical constants. The skin depth λ evidently provides a length scale, and consequently, its square divided by the thermal diffusivity $\kappa = K/\rho c$ is a unit of time. Further, if T_m denotes some characteristic temperature (such as the melting point) at which some desired change occurs, then, by inspection, we see that $(T_m - T_\infty)K/\lambda$ has units of flux. Introducing, therefore, the new independent variables

$$\xi = x/\lambda, \qquad \tau = \kappa t/\lambda^2, \tag{7.4a,b}$$

and the dimensionless temperature and flux

$$u(\xi,\tau) = \frac{T(x,t) - T_\infty}{T_m - T_\infty}, \qquad \varphi(\tau) = \frac{\lambda(1-R)}{K(T_m - T_\infty)}F_0(t), \tag{7.5a,b}$$

we transform Eq. (7.2) into the convenient dimensionless form

$$\frac{\partial u}{\partial \tau} = \frac{\partial^2 u}{\partial \xi^2} + \varphi e^{-\xi}. \tag{7.6a}$$

This we must solve under the conditions (7.3)

---

† Our former input flux F is exactly the unreflected portion $(1-R)F_0$ considered here.

$$u(\xi,0) = 0, \tag{7.6b}$$

$$u(\infty,\tau) = u_\xi(0,\tau) = 0, \tag{7.6c}$$

expressed now in the new variables.

Taking the Laplace transform of Eq. (7.6a), defining $\bar{u}(\xi,p) = \mathscr{L}\{u(\xi,\tau)\}$ and $\bar{\varphi}(p) = \mathscr{L}\{\varphi(\tau)\}$ according to the definition (6.1), and using the linearity and differentiation properties (6.6) and (6.11), we get the ordinary linear differential equation

$$p\bar{u} = d^2\bar{u}/d\xi^2 + \bar{\varphi}e^{-\xi} \tag{7.6d}$$

because the initial condition (7.6b) vanishes. Its solution, under the boundary conditions (7.6c), is easily seen to be

$$\bar{u}(\xi,p) = \frac{\bar{\varphi}(p)}{p-1}\left[e^{-\xi} - \frac{e^{-q\xi}}{q}\right], \quad \text{where} \quad q = \sqrt{p}, \tag{7.7}$$

which we must invert to recover the temperature response of the irradiated material. A formal inversion is not hard to obtain through the convolution theorem (6.35). In fact, each term of the image temperature (7.7) is available in Table 6.1 (entries 3 and 12) because

$$\frac{1}{p-1} = \frac{1}{q^2-1} = \frac{1}{2}\left(\frac{1}{q-1} - \frac{1}{q+1}\right).$$

If, however, we are only interested in the surface temperature T_0 (or its dimensionless embodiment u_0) rather than in the whole temperature distribution, then it suffices to evaluate (7.7) for $x = 0$. We then have

$$\bar{u}_0(p) \equiv \bar{u}(0,p) = \bar{\varphi}(p)\frac{1}{q(q+1)}, \tag{7.8}$$

which is easily inverted to give

$$u_0(\tau) = \int_0^\tau \varphi(\sigma)e^{(\tau-\sigma)}\text{erfc}\sqrt{\tau-\sigma}\,d\sigma, \tag{7.9a}$$

or, reverting to the dimensional variables (7.4),

$$T_0(t) = T_\infty + \frac{1-R}{\rho c \lambda} \int_0^t F_0(s) e^{(\kappa t - s)/\lambda^2} \mathrm{erfc} \frac{\sqrt{\kappa(t-s)}}{\lambda} \, ds. \qquad (7.9b)$$

There are two points to bear in mind regarding this last result. First, we evaluated the surface temperature *before* Laplace inversion; second, it is valid for *arbitrary* laser pulses. Indeed, the coefficient of $\bar{\varphi}$ in the transformed solution (7.7) is the problem's (transformed) Green's function $\bar{G}(\xi,p)$, and we can write formally $u(\xi,\tau) = \varphi(\tau) \star G(\xi,\tau)$. This Green's function, in fact, contains *all* the information on the temperature's spatial distribution. [The reader should note the similarities and differences between this result and the problem leading to the integral (4.5).] We end this section with general remarks about short- and long-time behavior, and we defer to the next section some detailed calculations for specific laser pulse shapes.

The behavior for short times is easy to obtain from the Laplace transform (7.8) if we remember the discussion of Sect. 6.6. For $p \to \infty$ we get $\bar{u}_0 \sim \bar{\varphi}/p$. Consequently, with the integration property (6.15), we have

$$u_0(\tau) \cong \int_0^\tau \varphi(\tau) \, d\tau, \qquad (7.10a)$$

or in dimensional form

$$T_0(t) \cong T_\infty + \frac{1-R}{\lambda \rho c} \int_0^t F_0(s) \, ds. \qquad (7.10b)$$

In other words, the material system responds initially by *integrating* the laser pulse. Physically, in the beginning, essentially all the laser's energy goes into heating a surface layer of depth λ, while the rest of the sample appears adiabatically insulated from this surface layer. Indeed, neither thermal conductivity nor diffusivity appears in Eq. (7.10b). This result can be also verified directly from Eqs. (7.9) because $e^z \mathrm{erfc}\sqrt{z} \cong 1 - 2(z/\pi)^{1/2}$ by virtue of the expansion property (d) of Sect. (4.1) and the definition (4.10b). Consequently, "short" times mean that $\tau^{1/2}$ must be much less than unity, *i.e.*, the diffusion length must be smaller than the skin depth. Taking $\kappa \cong 0.2 \text{ cm}^2/\text{s}$ (for Si at 800°C), and $\lambda \cong 1 \ \mu\text{m}$ (for an argon-ion laser), we find from Eq. (7.4b) that t must be much less than 50 nanoseconds for us to observe initial transient behavior.

Assume, on the other hand, that the laser had been switched on for only a duration t_1 (the details of the pulse's shape are irrelevant). We ask what is the surface temperature long after the pulse has been switched off. By the mean-value theorem, there exists a time $t^* \in [0,t_1]$ (and corresponding scaled times τ) such that formula (7.9a) can be written

$$u_0(\tau) = \varphi(\tau^*) \int_0^{\tau_1} e^{(\tau-\sigma)} \text{erfc}\sqrt{\tau-\sigma}\, d\sigma$$

$$= \varphi(\tau^*) \int_{\tau-\tau_1}^{\tau} e^{\sigma} \text{erfc}\sqrt{\sigma}\, d\sigma, \tag{7.11}$$

where the second equality follows from a simple change of variables. Since we are interested in the behavior for $t \gg t_1$, it suffices to examine the asymptotic behavior of the integrand, which, by virtue of Eqs. (4.11), is asymptotically equal to $(\pi\sigma)^{-1/2}$. Thus we have

$$u_0(\tau) \sim \varphi(\tau^*) \int_{\tau-\tau_1}^{\tau} \frac{d\sigma}{\sqrt{\pi\sigma}} = \frac{2\varphi(\tau^*)}{\sqrt{\pi}}\left(\sqrt{\tau} - \sqrt{\tau-\tau_1}\right)$$

$$\sim \varphi(\tau^*)\tau_1/\sqrt{\pi\tau}, \tag{7.12a}$$

where, applying the binomial theorem to $\sqrt{\tau-\tau_1}$, the second equality holds when $\tau \gg \tau_1$. We also get its dimensional form

$$T_0(t) \sim T_\infty + \frac{(1-R)F_0(t^*)t_1}{K}\left(\frac{\kappa}{\pi t}\right)^{1/2}. \tag{7.12b}$$

Therefore, for long times, we find that the average *energy* $F_0(t^*)t_1$ contained in the laser pulse causes a temperature decay proportional to $t^{-1/2}$. Contrary to the short-time behavior (7.10b), this result is independent of skin depth, but it does depend on κ and K. Physically, this simply means that the laser's energy decays into the sample by diffusion and that all "memory" of this energy's provenance is lost after a sufficiently long time.

Exercise 7.2: *More asymptotics*
Show that the result (7.12b) can also be obtained directly from the Laplace transform (7.7). With the same parameters as in the case of short times, what does a "long" time mean? Carry out the

expansions (7.10) and (7.12) to higher orders. Can you discuss likewise these asymptotic limits of the full temperature distribution
$T(x,t)$? ∎

7.2 LASER PROCESSING: PULSE SHAPES

The formulas (7.9), it is recalled, hold for any pulse shape. In particular, if we have a cw laser, then the input flux is simply proportional to
the Heaviside step function (6.4a). We then have

$$F_0(t) = F_1 H(t),$$
                                                                (7.13a)

where F_1 is the constant flux delivered by the laser. Defining the
corresponding dimensionless flux φ_1, we obtain the surface temperature
rise

$$u_0(\tau) = \varphi_1 \int_0^\tau e^\sigma \mathrm{erfc}\sqrt{\sigma}\, d\sigma.$$
                                                                (7.13b)

But this integral can be evaluated explicitly by parts, because, using the
differentiation formula (c) of Sect. (4.2), we have

$$\frac{d}{dz}(e^z \mathrm{erfc}\sqrt{z}) = e^z \mathrm{erfc}\sqrt{z} - \frac{1}{\sqrt{\pi z}}.$$
                                                                (7.13c)

Therefore,

$$u_0(\tau) = \varphi_1 \left(e^\tau \mathrm{erfc}\sqrt{\tau} + 2\sqrt{\tau/\pi} - 1 \right),$$
                                                                (7.14a)

or, reverting to dimensional variables,

$$T_0(t) = T_\infty + \frac{\lambda(1-R)F_1}{K}$$

$$\times \left[e^{\kappa t/\lambda^2} \mathrm{erfc}\frac{\sqrt{\kappa t}}{\lambda} + \frac{2}{\lambda}\left(\frac{\kappa t}{\pi}\right)^{1/2} - 1 \right].$$
                                                                (7.14b)

We may now ask what relation exists between these formulas and our
earlier estimate (4.21e). The second remark that follows Eqs. (7.3)
suggests that we should indeed recover the heating problem (4.21) in
the approximation $\lambda \to 0$. This means that the scaled time τ [cf.
Eq. (7.4b)] is almost always very large, and we can use the asymptotic

expansion of the error function in Eqs. (7.14) to get

$$u_0(\tau) \sim \varphi_1 \left(\frac{1}{\sqrt{\pi\tau}} + 2\sqrt{\tau/\pi} - 1 \right), \tag{7.15a}$$

which, for $\tau \gtrsim 100$, reduces to

$$u_0(\tau) \sim \varphi_1 2\sqrt{\tau/\pi}. \tag{7.15b}$$

This is precisely our earlier result (4.21e) in dimensionless form. On the other hand, when $\tau \ll 1$ (*i.e.*, if either the processing time is short or the skin depth is very large), we have the initial behavior

$$u_0(\tau) \cong \varphi_1 \tau. \tag{7.16}$$

This follows either from the general result (7.10) or from a short-time expansion of the error function in the particular case (7.14).

The surface temperature (7.14) is plotted in Fig. 7.1 as a function of scaled time, together with its limiting behavior (7.16) and (7.15b) for short and long times, respectively. We see that the surface temperature is a monotonically increasing function of time, and that it scales linearly with the laser flux F_1. This figure can be read in yet another way. From the definition (7.5a) of the scaled temperature it follows that

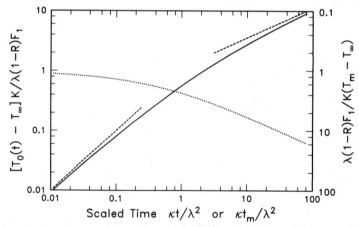

Fig. 7.1: Surface temperature (full line) as a function of time for the laser heating problem. The dashed lines correspond to the approximations (7.16) and (7.15b) for short and long times, respectively. The dotted line is the integrand of Eq. (7.13b). The use of the right-hand ordinate scale is explained in the text.

$$u_0(\tau_m) = 1 \tag{7.17a}$$

defines that time τ_m when the surface temperature will have first reached the melting or transition point T_m. Inserting this condition into Eq. (7.14a), we get

$$\varphi_1^{-1} = \left(e_m^{\tau}\operatorname{erfc}\sqrt{\tau_m} + 2\sqrt{\tau_m/\pi} - 1\right), \tag{7.17b}$$

and the log-log plot in Fig. 7.1 can be read "backward": A given flux F_1, if properly scaled (as shown on the right-hand ordinate), will produce a surface temperature T_m in a time t_m that can be read off directly from the abscissa.

Exercise 7.3: Power required for melting

Using the analytic expressions (7.15-16), estimate the time t_m as a function of laser power that is required to achieve a phase transformation. Carefully state the limits of validity of your estimates. ∎

Next in order of complexity, we can compute the effect of the rectangular laser pulse shown in Fig. 2.2. It is constant in the interval $(0, t_1)$, and it vanishes outside this interval. This is evidently an idealization of real pulses, because rise rates cannot be infinitely large. This example, however, illustrates the principle behind any analytic evaluation of temperature response. We simply note that the pulse is the sum of two Heaviside functions. Thus,

$$\varphi(\tau) = \varphi_1[H(\tau) - H(\tau-\tau_1)], \tag{7.18}$$

and the required surface temperature (7.9) can be expressed in terms of our previous result (7.14). Indeed, define by

$$U(t) = \begin{cases} 0 & \text{if } \tau < 0, \\ e^{\tau}\operatorname{erfc}\sqrt{\tau} + 2\sqrt{\tau/\pi} - 1 & \text{if } \tau \geq 0, \end{cases} \tag{7.19a}$$

the effect of a Heaviside pulse. Then, by the linearity of the integral (7.9a), we must have the surface response

$$u_0(\tau) = \varphi_1[U(\tau) - U(\tau-\tau_1)]. \tag{7.19b}$$

This is the sum of two shifted functions (7.19a), each of which picks up at a break point of the laser's flux schedule (7.18).

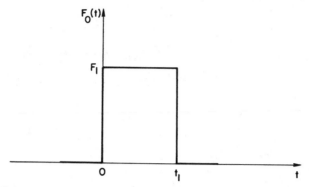

Fig 7.2: Rectangular laser pulse of duration t_1 and of intensity F_1.

Exercise 7.4: *Calculation using the Laplace transform*
Obtain this last result (7.19) from the general Laplace transform (7.8). ∎

Figure 7.3 displays the surface temperature as a function of time, scaled to the rectangular pulse length t_1. We see that the temperature rises as long as the laser is on, but that it drops immediately after the power drops to zero. Thus, the maximum temperature T_{max} occurs exactly at the pulse's end, $t = t_1$. This can be easily proved analytically, because the derivative of the U-function (7.19a) is $e^\tau \mathrm{erfc}\sqrt{\tau}$ [*cf.* Eq. (7.13b)], which never vanishes.† Consequently, similarly to the previous case (7.17), $u_0 \geq 1$ implies the condition

$$\varphi_1\left(e^{\tau_1}\mathrm{erfc}\sqrt{\tau_1} + 2\sqrt{\tau_1/\pi} - 1\right) \geq 1 \qquad (7.20)$$

for a phase transition to occur. This relation characterizes a region in the (t_1, F_1)-plane between the laser flux and its pulse length.

There are many other interesting questions about heat and mass transport, induced by laser irradiation. For example, we have assumed that the flux was constant in the *lateral* dimensions, a generally bad assumption. Moreover, cw lasers (or other sources of heat) often scan

---

† One shows that there is no maximum in the *open* intervals $(0, t_1)$ and (t_1, ∞). Therefore, T_{max} occurs exactly at $t = t_1$. For more smoothly varying laser schedules (even triangular ones), one shows that T_{max} always occurs *after* the maximum of the laser pulse.

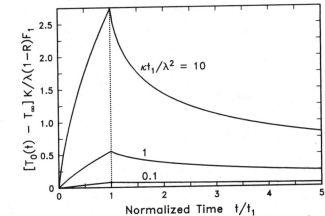

Fig. 7.3: Surface temperature response to a rectangular laser pulse. Note that this system, like all diffusion systems, has no inertia, *i.e.*, the temperature drops in phase with the laser power.

the material. Consequently, we often have moving sources, and these have gaussian radial shapes. These and other questions have been actively pursued [1,7-10].

7.3 THERMALLY STIMULATED DIFFUSION

Even linear diffusion problems may introduce variable diffusivities. In Sect. 4.5, we saw that programmed temperature changes induce additional driving forces on crystallizing systems. There, we had assumed that the diffusivity was temperature independent. This may be a good approximation in liquids as long as temperature excursions are moderate. In solids, however, diffusivities are often activated for reasons explained in Sect. 3.2. Consequently, they are strong functions of temperature and therefore of time. To motivate the following treatment of such cases, we turn to an old but popular method of surface analysis: *flash desorption.*

The method consists, essentially, in placing a clean sample in an evacuated ampule, where it is thoroughly degassed. A known amount of foreign gas is then introduced into the chamber, and it is allowed to equilibrate with the sample. Therefore, we know the initial adsorbed concentration $C_{s,e}$ of that foreign gas on our sample. If, now, the ampule is again evacuated, then that surface population is no longer in equilibrium with the gas phase, and desorption will occur. This can be monitored with pressure gauges or mass spectrometers, and the kinetics of desorption, *i.e.*, essentially the decay of the adsorbed concentration $C_s(t)$ can be followed. This method yields insight into the forces that bind molecules to surfaces. Nevertheless, its analysis depends on the solution of models for the desorption process [11,12]. It is generally assumed that we have a rate equation of the form

$$\dot{C}_s = -KC_s^\nu, \quad \text{with} \quad C_s(0) = C_{s,e}. \tag{7.21a}$$

Here, K and ν are the rate constant and the order of the (desorption) reaction, respectively. This differential equation is separable

$$dC_s/C_s^\nu = -K\,dt \tag{7.21b}$$

as long as K does not depend on concentration. If K is also time independent, then Eq. (7.21b) is easily solved:

$$Kt = \begin{cases} \ln(C_{s,e}/C_s) & \text{if } \nu = 1, \\ (\nu-1)\left(C_s^{1-\nu} - C_{s,e}^{1-\nu}\right) & \text{if } \nu \neq 1. \end{cases} \tag{7.22}$$

The important point to note is that the right-hand side of this equation consists of pure functions of C_s, while the left-hand side depends exclusively on time. Consequently, if K is any function of temperature T, and if the temperature is programmed to change, for example linearly†

$$T(t) = T_0 + \alpha t, \tag{7.23}$$

then the kinetics (7.22) are still true by merely substituting

$$\int_0^t K[T(s)]\,ds \tag{7.24}$$

for the left-hand side Kt, as should be clear from Eq. (7.21b).

---

† Since positive temperature ramps are more common in thermally stimulated processes, here we take the ramp rate α to be positive for heating. The opposite sign was used in Eq. (4.35a) for crystallization. Note that the following considerations are valid for *any* given temperature schedule $T(t)$.

The analysis of these kinetics is here not our concern. It suffices to indicate that there are, again, two competing forces: (i) a temperature ramp, which tends to *increase* K and, with it, the rate of desorption; (ii) the natural (isothermal) tendency of $\dot{C}_s$ to *decrease* with time according to Eqs. (7.21). Therefore, the desorption rate exhibits peaks, whose analysis provides information on surface binding and structure. This experimental method is called *thermal stimulation*. It has been applied to a wide variety of other physical systems, *e.g.*, thermally stimulated electronic or ionic currents in dielectrics and semiconductors [13].

Exactly the same situation holds for the diffusion equation (1.14) or (1.33) when the diffusivity depends on time through programmed temperature changes (7.23). If λ is some constant distance scale, then it is easy to see† that the change of variables

$$\xi = x/\lambda \quad \text{and} \quad \tau = \lambda^{-2} \int_0^t D[T(s)]\,ds \qquad (7.25\text{a,b})$$

transforms the diffusion equation (1.14) into $C_\tau = C_{\xi\xi}$, namely, an equation whose *coefficients are constant*. The same is evidently true in several space dimensions. Consequently, *any* solution obtained for constant D becomes formally valid by the mere substitution of ξ and τ for x and t. This method has been used to study impurity diffusion kinetics if the sample's surface does not accumulate these impurities [14-16]. The spectra exhibit peaks, just as in the case of flash desorption, because, on the one hand, D rises with temperature (thus causing larger fluxes), and on the other hand, the amount of diffusant in any finite sample decreases with time, which tends to decrease the surface flux.

Now, in particular, Eq. (7.25b) is a relation between physical time t and "stretched" time τ, and it is this relation we must explore. For definiteness, let us assume (*cf.* Chap. 3) that D has an Arrhenius dependence

$$D(T) = D_\infty e^{-E/kT} \qquad (7.26)$$

on T, where E and k are, as usual, an activation energy and Boltzmann's constant, respectively, and where D_∞ is a constant "pre-

---

† Just note that $\partial/\partial t = (D/\lambda^2)\,\partial/\partial\tau$.

exponential factor," which represents D at infinite temperature. Together with the ramp (7.23), its substitution into Eq. (7.25b) yields

$$\tau = (D_\infty/\lambda^2)\int_0^t e^{-E/k(T_0+\alpha s)}\,ds. \tag{7.27a}$$

The exponent shows, as in statistical mechanics, that E/kT is a natural variable for this problem. Therefore, the change of variables $s \to \beta = E/kT(s)$ transforms the integral into

$$\tau(t) = (ED_\infty)/(k\alpha\lambda^2)\int_{E/kT(t)}^{E/kT_0} e^{-\beta}\beta^{-2}\,d\beta. \tag{7.27b}$$

This integral, unfortunately, is not expressible in terms of elementary functions, but it is the difference between two integrals of the form

$$I(z) = \int_z^\infty e^{-\beta}\beta^{-2}\,d\beta, \tag{7.28a}$$

which is closely related to the *exponential integral* [17]. Thus, we have

$$\tau(t) = (ED_\infty)/(k\alpha\lambda^2)\{I[E/kT(t)] - I(E/kT_0)\}, \tag{7.29a}$$

and we are left wondering how to compute these integrals. In general, it is true, we must either compute the integrals numerically or use tables provided in Ref. [17]. On the other hand, unless E is very small, the argument E/kT (and *a fortiori* E/kT_0) is a very large number. For example, if the final temperature is $1000°C$ and if the activation energy is of the order of an eV, then that argument's value is approximately 10. Consequently, we are only interested in the asymptotic behavior of the integral (7.28a). Exactly as with the error function in Eqs. (4.11), repeated integration by parts produces the asymptotic expansion

$$I(z) = \frac{e^{-z}}{z^2}\left[1 - \frac{2}{z} + \frac{2\times3}{z^2} - \ldots\ldots\right], \tag{7.28b}$$

and we retain only the first term when the argument is positive and large. Therefore, the relation (7.29a) becomes asymptotically

$$\tau(t) \propto [E/kT(t)]^{-2}e^{-E/kT(t)} - [E/kT_0]^{-2}e^{-E/kT_0}. \tag{7.29b}$$

This function is plotted in Fig. 7.4 for various values of the initial energy ratio E/kT_0, together with an approximation, proved in the next

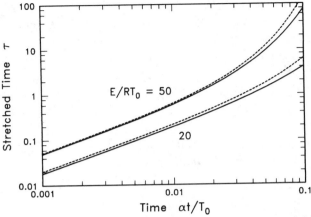

Fig. 7.4: The time-like (stretched) variable τ when the diffusivity depends explicitly on time.

exercise, that holds for moderate temperature excursions.

Exercise 7.5: *Short-time expansion*
Expand the exponent in Eq. (7.27a) and find an approximation valid for short times. Invert for $t(\tau)$ in that approximation. What is an appropriate length scale λ? ∎

7.4 GREEN'S FUNCTIONS AND THE USE OF INTEGRAL EQUATIONS

Green's functions are often cloaked in mystery, which is unfortunate because their motivation is so simple. Recall the formulas (1.29b), (4.5), (6.33c), (6.36), and (7.9). In all of these, solutions to given problems are expressed as convolution integrals over some *a priori arbitrary* condition. That condition might be an initial condition, as in Eqs. (1.29b) and (4.5), or an inhomogeneous forcing term, as in Eqs. (6.33c) and (7.9), or yet a Dirichlet condition, as in Eq. (6.36). The functions with which these conditions are convolved are precisely Green's functions. We have verified (often by their very construction) that each of these is a solution of the *same* underlying differential equation and that it satisfies some *homogeneous* auxiliary condition. For example, the function that multiplies the arbitrary surface concentration $C_0(t)$ in Eq. (6.36) is a solution of the diffusion equation, and it vanishes at both ends of the interval $[0,\infty]$. In general, each given problem, expressed through a differential equation *and* auxiliary conditions, has a Green's function G with these properties. A thorough investigation of Green's functions belongs in texts on differential equations. Here, we shall continue to construct them, almost automatically, through Laplace transforms. The following example illustrates the procedure.

Once again, consider impurity diffusion in a half space under the usual initial and far-field conditions

$$C(x,0) = C(\infty,t) = C_\infty. \tag{7.30a,b}$$

Rather than specify either concentration or flux at the surface $x = 0$, let us impose a boundary condition of the type

$$D\frac{\partial C}{\partial x} = k(C - C_e) \quad \text{at } x = 0, \tag{7.31}$$

where k is some positive number, and C_e is any given positive function of time. In other words, we specify a linear combination of flux and field at the boundary. This is called a *radiation boundary condition* because its prototype is to be found in Newton's law of cooling. It reduces to a Dirichlet condition $C_0 = C_e$ when the parameter k is very large, and to a (homogeneous) Neumann condition $J_0 = 0$ when $k \to 0$. The radiation boundary condition (7.31) is often used to simulate non-equilibrium conditions at surfaces. Its physical content is merely that the surface flux must accommodate any deviation from equilibrium at the surface, k being a "reaction constant." We have seen it in at least two different contexts: the boundary condition (2.9a) for oxidation kinetics, and the exchange-flux expression (2.28b) for the BCF theory.

The diffusion problem, as stated, is linear. It can be solved either directly or by straightforward application of the Laplace transform. But we would like this example to buy us a method of solution applicable to nonlinear boundary conditions as well. That application will be found in the next section. Here, we solve our problem by reducing it to the solution of an *integral equation*. The procedure is very simple. Rather than solve the stated problem, let us assume that the condition (7.31) is replaced by an arbitrary Neumann condition. In other words, we assume that the surface flux $-D\partial C/\partial x\,|_{x=0}$ is a given, but as yet unknown, function $J_0(t)$ of time.

The Laplace transform of the diffusion equation, under the conditions (7.30), yields

$$\overline{C}(x,p) = p^{-1}C_\infty + A e^{-x\sqrt{p/D}}, \tag{7.32a}$$

exactly as in Sect. 6.4. The constant A can be expressed in terms of this flux, because $\overline{J}_0 = -D\overline{C}'\,|_{x=0}$. Therefore, we have

$$\overline{C}(x,p) = \frac{C_\infty}{p} + \overline{J}_0\,\frac{e^{-x\sqrt{p/D}}}{\sqrt{pD}}. \tag{7.32b}$$

The first term's inverse transform is just the constant C_∞ because of entry 4 of Table 6.1, and the second term is the product of two factors. The second factor's inverse transform appears in entry 9 of that table, and, using the convolution theorem (6.35), we get

$$C(x,t) = C_\infty + \int_0^t J_0(s)\,\frac{e^{-x^2/4D(t-s)}}{\sqrt{\pi D(t-s)}}\,ds. \tag{7.33}$$

This formula expresses the concentration distribution essentially as a convolution of an arbitrary flux and the gaussian (4.3).† In other words, should we ever find the flux J_0, then the full concentration distribution would be completely determined by the single quadrature (7.33). In particular, setting $x = 0$, we obtain the surface concentration

$$C(0,t) = C_\infty + \int_0^t J_0(s)\,\frac{ds}{\sqrt{\pi D(t-s)}}. \tag{7.34}$$

Note that, up to now, the boundary condition (7.31) has not been used at all. If, however, we recognize that its left-hand side must be equal and opposite to J_0, then, inserting Eq. (7.34) into Eq. (7.31), we produce the formula

---

† By now, the reader surely knows that gaussians satisfy the diffusion equation and that their gradient vanishes at their maximum's position, *i.e.*, they are authentic Green's functions for this Neumann problem.

$$-J_0(t) = k\left[C_\infty - C_e + \int_0^t J_0(s) \frac{ds}{\sqrt{\pi D(t-s)}} \right]. \qquad (7.35)$$

This expression is called an integral equation, because the unknown function $J_0(t)$ appears both inside and outside an integral. In many ways, such expressions are the natural analog (for functions) of linear equations (for algebraic quantities). The method of solution that we shall explore in the next section will emphasize that analogy. There are many types of integral equations. Their classification is beyond our scope, but we notice that this one is linear in the unknown function and that the range of integration is limited to the interval $[0,t]$. Technically, it is called a "Volterra integral equation of the second kind, with weakly singular, displacement kernel" [18]. To elaborate somewhat, the function $(t-s)^{-1/2}$ that multiplies $J_0(s)$ is called the *kernel* of this integral equation. It is of *displacement* type because it depends exclusively on the retarded argument $t-s$. The kernel is called *weakly singular* because its singularity is integrable. This last remark is of some importance for any numerical method of solution, as we shall soon see. But are numerical methods required for this simple problem? One would hope not, because, as mentioned earlier, linear diffusion problems under radiation boundary conditions can be solved directly.

Indeed, because Eq. (7.35) is linear, and because of its displacement kernel, we can apply the Laplace transform and its convolution theorem. If the equilibrium concentration C_e is assumed constant, then

$$-\bar{J}_0 = k\left[p^{-1}(C_\infty - C_e) + \bar{J}_0(pD)^{-1/2} \right], \qquad (7.36a)$$

because $\mathcal{L}\{(\pi t)^{-1/2}\} = p^{-1/2}$. This is merely a linear equation in the transform $\bar{J}_0$, which is trivially solved:

$$\bar{J}_0(p) = -kD \frac{C_\infty - C_e}{\sqrt{Dp}\left(k + \sqrt{Dp}\right)}. \qquad (7.36b)$$

Its inversion is equally simple if we use entry 12 of Table 6.1 and the change of scale property (6.8), and we get the flux's time dependence

$$J_0(t) = -k(C_\infty - C_e)e^{k^2 t/D} \operatorname{erfc}(k\sqrt{t/D}). \qquad (7.37)$$

This procedure deserves a few remarks. First, it is clear that we have completely solved our diffusion problem, because, as mentioned earlier, insertion of the flux (7.37) into the Green's-function representation (7.33) of the concentration yields a closed-form expression for $C(x,t)$.

Exercise 7.6: *Concentration distribution under radiation boundary conditions*
Carry out the calculation just outlined. [Hint: Integrate by parts using the formula (7.13c).] ∎

Second, there are circumstances when C_e is a given function of time; the effect of temperature ramps, discussed in Sect. 4.5, is case in point. The previous problem can still be solved analytically, as is shown in the next

Exercise 7.7: *Exact solution with a variable forcing function*
Assume that the equilibrium concentration is a variable, but known, function of time. Solve the integral equation (7.35). ∎

Third and most important, we have, in effect, decomposed the given problem in the two independent variables, x and t, into two separate problems, each carrying over a *single* independent variable. Indeed, the integral equation (7.35) involves only the time t, and the quadrature (7.33) is again over the time-like variable s. Fourth, there exist circumstances when one may be only concerned with the surface flux or the surface concentration, because, for example, the whole concentration distribution is unobservable. Integral-equation methods then provide a direct route to the quantities of interest. Fifth, in contrast to partial differential equations, integral equations are very easily solved numerically with great accuracy, as we shall soon see. Finally, the reduction of a given problem to an integral equation often provides means of studying the analytic properties and asymptotic behavior of solutions [19-21].

Before going on to the final section of this book, let us note that we are already familiar, from our work with lasers, with the particular function of time (7.37). Indeed, this very function, albeit with a different time scaling, appears in Eqs. (7.9). It is even plotted in

Fig. 7.1 as a dotted line, and it would represent the surface temperature response to a laser pulse in the form of a delta function of time.

7.5 APPLICATION TO NONLINEAR SURFACE CONDITIONS

Are we to lose hope if faced with a problem where the boundary
condition (7.31) is *nonlinear*? For example, chemical reactions are
often *not* first order in the concentrations, and for thermal problems,
radiation involves the fourth power of the surface temperature. Indeed,
the study of chemical boundary layers [22] requires similar analyses.
Let us explore ways of solving the same diffusion problem as in the last
section, but where the boundary condition (7.31) is replaced by

$$D\frac{\partial C}{\partial x} = K(C - C_e)^\nu \quad \text{at } x = 0, \tag{7.38}$$

in which ν is an arbitrary power and the reaction constant is now called
K to distinguish it from the linear case. The method of integral equations leads us quickly to a formal solution, because the development
from Eqs. (7.32) to Eq. (7.34) is independent of this boundary condition. Consequently, it suffices merely to replace Eq. (7.35) by

$$-J_0(t) = K\left[C_\infty - C_e + \int_0^t J_0(s)\frac{ds}{\sqrt{\pi D(t-s)}} \right]^\nu. \tag{7.39}$$

It is this nonlinear integral equation that we shall solve numerically.
Once this is accomplished, then, as before, the complete concentration
distribution results from the quadrature (7.33), which now must be
performed numerically.

Let us first reduce the integral equation to a standard dimensionless
form. Taking a cue from the solution (7.37) of the linear radiation
problem, we see that $\tau = k^2 t/D$ and $J_0/k(C_\infty - C_e)$ are dimensionless
because $[k] = \text{cm/s}$. By inspection, we see that $K(C_\infty - C_e)^{\nu-1}$ also
has dimensions of velocity. Consequently, if we define a scaled time
and flux

$$\tau = K^2(C_\infty - C_e)^{2(\nu-1)}t/D, \quad \varphi(\tau) = -J_0(t)/K(C_\infty - C_e)^\nu \tag{7.40a,b}$$

and if we take the νth root of Eq. (7.39), then the integral equation
takes the dimensionless form

$$[\varphi(\tau)]^{1/\nu} = 1 - \int_0^\tau \varphi(\sigma) \frac{d\sigma}{\sqrt{\pi(\tau-\sigma)}}. \tag{7.41}$$

It is just as easy to solve this integral equation when the constant, unity, on its right-hand side is replaced by an arbitrary, but known forcing function f of time (*cf*. Exercise 7.7 above). Consequently, let us discuss the solutions, numerical and analytic, of the following standard integral equation

$$[\varphi(t)]^{1/\nu} = f(t) - \int_0^t \varphi(s) \frac{ds}{\sqrt{\pi(t-s)}}, \tag{7.42}$$

where, for purely typographical reasons, we now call the dimensionless independent variables t and s.

The numerical solution of integral equations depends on methods for estimating the integral. In other words, we need a suitable quadrature formula. Let us seek a solution if $t \in [0,T]$, where T is any pre-assigned final value of t. If

$$t_0 = 0, t_1, ..., t_n, ..., t_N = T \tag{7.43a}$$

is any subdivision of this interval, then, just as in our discussion in Sect. 1.4, "seeking a solution" means that we wish to find the values of the unknown function

$$\varphi(t_n) \equiv \varphi_n \tag{7.43b}$$

at these nodes. At first glance, we might think that even the trapezoidal rule for computing integrals would be adequate. A moment's reflection, however, shows that such a procedure cannot succeed, because the kernel is singular whenever the integration variable s passes through any of the nodes (7.43a). If, on the other hand, we are able to represent the unknown function $\varphi(t)$ as a linear combination of simple integrable functions — call them $S_n(t)$ — and if these basis functions select the proper values φ_n at the nodes t_n, then our problem will be solved, in principle. Therefore, let us write

$$\varphi(t) = \sum_{n=0}^N \varphi_n S_n(t), \tag{7.44}$$

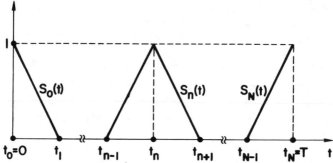

Fig. 7.5: Tent functions (finite elements) for the solution of integral equations.

where the S's are "tents," centered at each node and of unit height. They are represented in Fig. 7.5, with the understanding that each $S_n(t)$ vanishes identically outside the interval $[t_{n-1}, t_{n+1}]$. Therefore, the sum (7.44) clearly satisfies the "collocation" condition (7.43b) at each of the nodes (7.43a). Indeed, that sum is simply a linear interpolation of adjacent functional values of φ. Introducing Eq. (7.44) into the integral of Eq. (7.42) and expressing that the integral equation must hold at any node t_n, we obtain

$$\varphi_n^{1/\nu} = f_n - \sum_{m=0}^{N} A_{nm} \varphi_m \qquad (n = 0, \ldots, N), \tag{7.45a}$$

where, similarly to Eq. (7.43b), the f_n are the node values of $f(t)$, and where the matrix elements are

$$A_{nm} = \int_0^{t_n} S_m(s) \frac{ds}{\sqrt{\pi(t_n - s)}}. \tag{7.45b}$$

This is an $(N+1) \times (N+1)$ system of *algebraic* equations in the unknown functional values φ_n. To solve this system, we need expressions for the matrix elements (7.45b), and to perform these integrals, we now need analytic expressions for the basis functions.

First, notice that the $N+1$ nodes (7.43a) are not necessarily equidistant. The following calculations, however, are easiest if we assume equidistance, namely,

$$t_n = n\Delta t \qquad (n = 0, \ldots, N), \quad \text{where} \quad \Delta t = T/N. \tag{7.46}$$

Since each basis function is piecewise linear, we easily get the formulas

$$S_0(t) = \begin{cases} 1 - t/\Delta t & 0 \le t \le t_1, \\ 0 & \text{elsewhere,} \end{cases} \tag{7.47a}$$

$$S_n(t) = \begin{cases} 1 - |t - t_n| / \Delta t & t_{n-1} \le t \le t_{n+1} \quad (n = 1,...,N-1), \\ 0 & \text{elsewhere,} \end{cases} \tag{7.47b}$$

$$S_N(t) = \begin{cases} 1 + (t - T)/\Delta t & t_{N-1} \le t \le T, \\ 0 & \text{elsewhere.} \end{cases} \tag{7.47c}$$

The integrals (7.45b) are now convergent. They can be performed explicitly, and we find

$$A_{00} = 0, \tag{7.48a}$$

$$A_{nm} = 0 \quad \text{if} \quad m > n, \tag{7.48b}$$

$$A_{nn} = (4/3)\sqrt{\Delta t/\pi} \quad (n = 1,....,N), \tag{7.48c}$$

$$A_{n0}/A_{nn} = (n-1)^{3/2} - n^{1/2}(n-3/2) \quad (n = 1,....,N), \tag{7.48d}$$

$$A_{nm}/A_{nn} = (n-m-1)^{3/2} + (n-m+1)^{3/2} - 2(n-m)^{3/2}$$

$$\text{if} \quad 0 < m < n = 2,....,N. \tag{7.48e}$$

After this algebraic spell, let us attempt to interpret these results. First, Eq. (7.48a) is obvious because we are integrating over an interval of length zero. Second, Eq. (7.48b) also makes sense because we are integrating a null function. Therefore, the matrix A_{nn} is triangular with zero entries above its principal diagonal. This is a feature of all Volterra integral equations. Third, all elements (7.48c) of the principal diagonal are equal, except the first, which is zero. Finally, all the remaining matrix elements depend exclusively on the difference $n-m$, because the kernel itself depends on the difference $t-s$. Thus, all elements of a given subdiagonal are equal, except the first, A_{n0}, which is given by Eq. (7.48d). It follows that all the information about this matrix resides in its two first columns, A_{n0} and A_{n1}.

With this information, we can rewrite the system (7.45a) in the form

$$\varphi_0^{1/\nu} = f_0, \qquad\qquad (7.49a)$$

$$\varphi_n^{1/\nu} + A_{nn}\varphi_n = f_n - A_{n0}\varphi_0 - \sum_{m=1}^{n-1} A_{n-m+1,1}\varphi_m$$

$$(n = 1,\dots\dots,N), \qquad (7.49b)$$

where the first and last terms have been extracted from the sum. This is a recursion whose first member, explicitly given by Eq. (7.49a), is simply $\varphi_0 = f_0^\nu$. The remaining equations (7.49b) are algebraic equations in the current unknown φ_n in terms of a right-hand side that is a linear combination of all the previous φ's. Therefore, each step of this recursion involves, at most, the solution of a single algebraic equation, and this is very easily handled by standard numerical root-finding algorithms such as Newton's. If the exponent ν is either 2 or 0.5, then the φ's are merely solutions of quadratic equations. Equations (7.49) are evidently linear when $\nu = 1$. The solutions of Eq. (7.41) when the forcing function f is unity are shown in Fig. 7.6 for the exponents $\nu = 0.5$, 1, and 2. The case $\nu = 1$, we have seen in the last section, has the analytic solution $e^t \mathrm{erfc}\sqrt{t}$. This provides a means to test the accuracy of this numerical method. If Δt is of the order 10^{-2}, then the numerical solution has at least three correct significant digits, except for the first two time steps.

The numerical method (7.49) rests crucially on the representation (7.44) of the unknown function in terms of very simple basis functions $S_n(t)$. As noted, this representation is, in fact, a linear interpolation between nodes. By choosing other basis functions (also called "finite elements"), one can obtain higher-order schemes that are also more accurate [23,24]. Indeed, the tent functions (7.47) are continuous but not continuously differentiable, so that formula (7.44) would be inadequate if estimates of derivatives of $\varphi(t)$ where needed. The numerical scheme (7.49) is nonetheless very powerful because it holds, almost unchanged, for nonlinearities other than powers. In fact, if the left-hand side of the integral equation (7.42) is replaced by an *arbitrary* function F of φ, *i.e.*, if we wish to solve

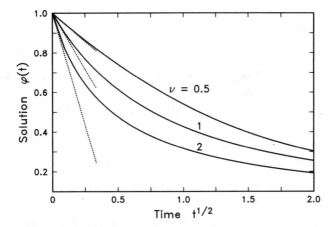

Fig. 7.6: Numerical solutions, for various values of ν, of the nonlinear integral equation (7.41). Dashed lines represent the starting solution (7.54).

$$F[\varphi(t)] = f(t) - \int_0^t \varphi(s)\frac{ds}{\sqrt{\pi(t-s)}}, \qquad (7.50)$$

then it suffices to replace the left-hand sides of Eqs. (7.49) by the functional values $F(\varphi_n)$ at the nodes (7.46).

7.6 STARTING SOLUTIONS AND AN EPILOGUE

There is a last question of some interest in connection with the nonlinear equation (7.42). We might only be interested in the unknown function's behavior for small values of time. This would be the case when the time constant [*cf.* Eq. (7.40a)] of a given physical system is very large. Alternatively, the forcing function $f(t)$ may have very rapid rates of change around $t = 0$, and any reasonable subdivision of time (7.46) would be inadequate. We will now show that it is very easy to find the analytic behavior around the origin.

Assume that the forcing function has a known expansion

---

† Here, the index n labels the various powers of $\sqrt{t}$. It must not be confused with that label for nodes in the last section. The odd coefficients f_{2n+1} would all vanish for functions analytic in t.

$$f(t) = \sum_{n=0}^{\infty} f_n t^{n/2} \tag{7.51a}$$

in powers of $\sqrt{t}$.† The constant term f_0 must not vanish, as will soon be clear. We then seek a similar expression for $\varphi(t)$. If it exists, then its arbitrary power $\psi \equiv \varphi^{1/\nu}$ is similarly expressible. Thus, we write

$$\varphi(t) = \sum_{n=0}^{\infty} \varphi_n t^{n/2}, \qquad \psi(t) \equiv [\varphi(t)]^{1/\nu} = \sum_{n=0}^{\infty} \psi_n t^{n/2}, \tag{7.51b,c}$$

and our problem reduces to the determination of the coefficients φ_n.

Inserting expansions (7.51) into the integral equation (7.42), performing the integrals using the convolution theorem (6.35) and the seventh entry of Table 6.1 [this is equivalent to the calculation (6.37) of the beta function], and identifying equal powers of $\sqrt{t}$, we produce the recursion

$$\psi_0 = f_0 = \varphi_0^{1/\nu}, \tag{7.52a}$$

$$\psi_n = f_n - c_{n-1}\varphi_{n-1}, \qquad n \geq 1. \tag{7.52b}$$

Here, the second equality (7.52a) follows from the continuity of f, φ, and ψ around the origin, and the coefficients c_n are

$$c_n = \frac{\Gamma(1 + n/2)}{\Gamma(3/2 + n/2)}. \tag{7.52c}$$

Exercise 7.8: *Relations for the c's*
Verify that $c_0 = 2/\sqrt{\pi}$ and that the coefficients (7.52c) obey the recursions

$$c_n c_{n-1} = 2/(n + 1) \qquad \text{and} \qquad c_n = n c_{n-2}/(n + 1).$$

Hence, compute the first few c's. ∎

It remains to express the coefficients of ψ in terms of those of φ. Putting momentarily $\mu = 1/\nu$ for convenience, application of the binomial theorem yields

$$\psi(t) = [\varphi_0 + \varphi_1\sqrt{t} + O(t)]^\mu$$

$$= \varphi_0^\mu [1 + \mu(\varphi_1/\varphi_0)\sqrt{t}] + O(t) \tag{7.53a}$$

to first order in $\sqrt{t}$. Consequently, we have ψ's first two coefficients

$$\psi_0 = \varphi_0^\mu, \quad \psi_1/\psi_0 = \mu\varphi_1/\varphi_0, \tag{7.53b}$$

which, when inserted into the recursion (7.52), produce the solution

$$\varphi_0 = f_0^\nu, \quad \varphi_1/\varphi_0 = \nu f_0^{-1}(f_1 - c_0\varphi_0). \tag{7.54a}$$

This first-order starting solution

$$\varphi(t) = \varphi_0 + \varphi_1\sqrt{t} \tag{7.54b}$$

is drawn dashed in Fig. 7.6.

The computations illustrated in Eqs. (7.53) can evidently be carried out to any order. The reader is encouraged to work out the second-order terms to feel the need for more efficient tools than binomial expansions. Fortunately, there exist formal procedures to find arbitrary powers of power series [25,26], and these investigations perhaps deserve more attention from scientists curious about diffusion processes. Finally, we are not limited to powers, and the nonlinear generalization (7.50), too, has starting solutions as shown in the last

Exercise 7.9: *Starting solutions for Eq. (7.50)*
Find a procedure to calculate the coefficients ψ_n for general nonlinear functions $\psi = F[\varphi]$. Hence, find the second-order starting solution of the integral equation (7.50). [Hint: Calculate the successive derivatives of ψ.] ∎

The last few sections of this book are paradigmatic, for they show the wide variety of methods that can be brought to bear on particular diffusion problems. Furthermore, they show that it is quite artificial to speak of numerical *vs.* analytic methods. Indeed, in this instance, analysis allows the reduction of a nonlinear problem to an integral equation. For this, a clear understanding of Laplace transforms and of Green's functions is necessary. On the other hand, numerics are required to find solutions of this integral equation. Then, to come full circle, analysis permits the establishment of starting solutions that are often useful, if not vital. Consequently, analytic and numerical methods appear as complementary tools for the solution of diffusion problems. An introduction to, and the interplay of these methods was my intention.

In this book I have also tried to indicate that the solution of a given diffusion problem, like any problem, requires a firm grasp of its physico-chemical background. Models abound, but few survive, often because some basic principle, such as conservation, is violated. And I would be content if only that message were carried.

REFERENCES

1. L.G. Pittaway, "The Temperature Distributions in Thin Foil and Semi-infinite Targets Bombarded by an Electron Beam," *Brit. J. Appl. Phys.* **15**, 967 (1964).

2. D. Maydan, "Micromachining and Image Recording on Thin Films by Laser Beams," *Bell Syst. Tech. J.* **50**, 1761 (1971).

3. R.J. von Gutfeld and P. Chaudhari, "Laser Writing and Erasing on Chalcogenide Films," *J. Appl. Phys.* **43**, 4688 (1972).

4. R.A. Ghez and R.A. Laff, "Laser Heating and Melting of Thin Films on Low-Conductivity Substrates," *J. Appl. Phys.* **46**, 2103 (1975).

5. R. Srinivasan, "Ablation of Polymers and Biological Tissue by Ultraviolet Lasers," *Science* **234**, 559 (1986).

6. J. Marshall, S. Trokel, S. Rothery, and R.P. Krueger, "Photoablative Reprofiling of the Cornea using an Eximer Laser," *Lasers in Ophthalmology* **1**, 21 (1986).

7. I.W. Boyd and J.I.B. Wilson, "Laser Processing of Silicon," *Nature* **303**, 481 (1983).

8. I.B. Khaibullin and L.S. Smirnov, "Pulsed Annealing of Semiconductors. Status Report and Unsolved Problems," *Fiz. Tekh. Poluprovodn.* **19**, 569 (1985). [Engl. transl. in *Sov. Phys. Semicond.* **19**, 353 (1985).]

9. D. Rosenthal, "The Theory of Moving Sources of Heat and its Application to Metal Treatments," *Trans. ASME* **68**, 849 (1947).

10. H.E. Cline and T.R. Anthony, "Heat Treating and Melting with a Scanning Laser or Electron Beam," *J. Appl. Phys.* **48**, 3895 (1977).

11. G. Ehrlich, "Kinetic and Experimental Basis of Flash Desorption," *J. Appl. Phys.* **32**, 4 (1961).

12. P.A. Redhead, "Thermal Desorption of Gases," *Vacuum* **12**, 203 (1962).

13. *Thermally Stimulated Relaxation in Solids*, P. Bräunlich, Ed. (Springer-Verlag, Berlin 1979).

14. G. Carter, B.J. Evans, and G. Farrell, "Gas Evolution from a Solid following De-trapping and Diffusion during Tempering," *Vacuum* **25**, 197 (1975).

15. M.A. Frisch, "Studies on Non-equilibrium Reactions by Knudsen Effusion Mass Spectrometry," in *Adv. in Mass Spectrometry* **8A**, pp. 391-401, A. Quayle, Ed. (Heyden, London 1980).

16. J.D. Fehribach, R. Ghez, and G.S. Oehrlein, "Asymptotic Estimates of Diffusion Times for Rapid Thermal Annealing," *Appl. Phys. Lett.* **46**, 433 (1985).

17. M. Abramowitz and I.A. Stegun, *Handbook of Mathematical Functions*, Chap. 3 (reprinted by Dover, New York 1965).

18. F.G. Tricomi, *Integral Equations* (Interscience, New York 1957). [Also reprinted by Dover, New York 1986.]

19. W.R. Mann and F. Wolf, "Heat Transfer between Solids and Gasses under Nonlinear Boundary Conditions," *Quart. Appl. Math.* **9**, 163 (1951).

20. R. Ghez and J.S. Lew, "Interface Kinetics and Crystal Growth under Conditions of Constant Cooling Rate," *J. Cryst. Growth* **20**, 273 (1973).

21. W.E. Olmstead and R.A. Handelsman, "Diffusion in a Semi-infinite Region with Nonlinear Surface Dissipation," *SIAM Review* **18**, 275 (1976).

22. P.L. Chambré and A. Acrivos, "On Chemical Surface Reactions in Laminar Boundary Layers," *J. Appl. Phys.* **27**, 1322 (1956).

23. C. Wagner, "On the Numerical Solution of Volterra Integral Equations," *J. Math. & Phys.* **32**, 289 (1954).

24. P. Linz, "Numerical Methods for Volterra Integral Equations with Singular Kernels," *SIAM J. Numer. Anal.* **6**, 365 (1969).

25. J.C. Jaeger, "Conduction of Heat in a Solid with a Power Law of Heat Transfer at its Surface," *Proc. Cambr. Phil. Soc.* **46**, 634 (1950).

26. A.C. Norman, "Computing with Formal Power Series," *ACM Trans. Math. Software* **1**, 346 (1975).

APPENDICES

A: RANDOM WALKS IN HIGHER DIMENSIONS

The considerations of Sects. 1.1-2 are here extended to higher dimensions. Particular attention is given to the notion of flux. Although, for simplicity, we emphasize derivations in *two* space dimensions, these considerations carry over to any number of dimensions. Some familiarity with tensor algebra [1,2] would be helpful.

Assume that a list of sites requires *pairs* (i,j) of integers and that $N_{i,j}$ represents the number of particles at the subscripted site. Assume also that *two* distinct transition frequencies Γ_1 and Γ_2 are required to describe jumps along the ith and jth directions, respectively. The two-dimensional lattice of integer points is represented in Fig. A.1, and, exactly as in Eqs. (1.1-3) of Sect. 1.1, we find that the particle flows in either direction are

$$J_{i+1/2,j} = \frac{1}{2}\Gamma_1(N_{i,j} - N_{i+1,j}),$$

$$J_{i-1/2,j} = \frac{1}{2}\Gamma_1(N_{i-1,j} - N_{i,j}), \qquad \text{(A.1a)}$$

and

$$J_{i,j+1/2} = \frac{1}{2}\Gamma_2(N_{i,j} - N_{i,j+1}),$$

$$J_{i,j-1/2} = \frac{1}{2}\Gamma_2(N_{i,j-1} - N_{i,j}). \qquad \text{(A.1b)}$$

The particle distribution's time variation is

$$\dot{N}_{i,j} = \frac{1}{2}\Gamma_1(N_{i+1,j} + N_{i-1,j} - 2N_{i,j})$$

$$+ \frac{1}{2}\Gamma_2(N_{i,j+1} + N_{i,j-1} - 2N_{i,j}), \qquad \text{(A.2)}$$

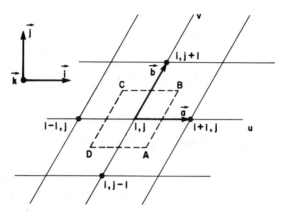

Fig. A.1: Oblique lattice for the 2-d random walk.

which can also be expressed in "divergence" form

$$\dot{N}_{i,j} = -(J_{i+1/2,j} - J_{i-1/2,j}) - (J_{i,j+1/2} - J_{i,j-1/2}). \qquad (A.3)$$

To find the continuum form of these equations, we must take coordinate transformations somewhat seriously. The integer sites (i,j) can be thought of either as integer points of the real R^2-plane (u,v) or as specified points (x_i, y_j) of our euclidean space E^2. The specification follows if we impose a one-to-one mapping

$$x = x(u,v), \qquad y = y(u,v) \qquad (A.4)$$

between arbitrary points of these two spaces. Because E^2 is endowed with an orthonormal base $(\mathbf{i}, \mathbf{j})$ (and thus with a metric), it follows that an arbitrary point has the vector representation $\mathbf{x} = x\mathbf{i} + y\mathbf{j}$, such that $x = \mathbf{x} \cdot \mathbf{i}$ and $y = \mathbf{x} \cdot \mathbf{j}$. In view of the mapping (A.4), its differential takes either one of two forms

$$d\mathbf{x} = \mathbf{i}\,dx + \mathbf{j}\,dy = (x_u\mathbf{i} + y_u\mathbf{j})\,du + (x_v\mathbf{i} + y_v\mathbf{j})\,dv, \qquad (A.5)$$

and the coefficients of du and dv define new basis vectors $\mathbf{a}$ and $\mathbf{b}$, tangent to the coordinate lines $u = $ const and $v = $ const, which are generally neither orthogonal nor of unit length, nor yet of constant length as one travels along coordinate lines. If, however, the mapping (A.4) is linear (thus with constant coefficients), then $\mathbf{a}$ and $\mathbf{b}$ maintain constant length, and we have defined a (generally oblique) the scalar products

$$\frac{\partial x^\beta}{\partial u^\alpha} = \mathbf{a}_\alpha \cdot \mathbf{e}_\beta \tag{A.10}$$

in the euclidean metric $\mathbf{e}_\alpha \cdot \mathbf{e}_\beta = \delta_{\alpha\beta}$, and they are constants — call them A^β_α — if the mapping is linear, as will henceforth be assumed. Recall, finally, that an arbitrary vector field $\mathbf{V}$ of E^2 has the two decompositions

$$\mathbf{V} = V^\alpha(x)\mathbf{e}_\alpha = V^\alpha(u)\mathbf{a}_\alpha, \tag{A.11a}$$

which define its (contravariant) components in each basis, and, with the transformation law (A.9b) for basis vectors, we get the transformation law

$$V^\alpha(x) = \frac{\partial x^\alpha}{\partial u^\beta} V^\beta(u) = A^\alpha_\beta V^\beta(u) \tag{A.11b}$$

for components. We will not dwell further on notions of covariance, metric tensor, and dual basis, because they are not needed in the present context of rectilinear lattices. They are required, however, for a study of hopping on *strained* lattices.

Returning to random walks, Eq. (A.7) can be written

$$\frac{\partial \hat{N}}{\partial t} = D^{\alpha\beta}(u)\frac{\partial^2 \hat{N}}{\partial u^\alpha \partial u^\beta} \tag{A.12a}$$

in the condensed notation, where

$$D^{\alpha\beta}(u) = \frac{1}{2}\begin{pmatrix} \Gamma_1 & 0 \\ 0 & \Gamma_2 \end{pmatrix} \tag{A.12b}$$

is a diagonal matrix formed from the transition frequencies. We will now see that it is also a second-order tensor, the *diffusivity tensor* $\mathbf{D}$, whose components in the coordinates $(u) \equiv (u^1, u^2)$ are given precisely by Eq. (A.12b). Note, first, that Eq. (A.12a) can also be written in divergence form

$$\frac{\partial \hat{N}}{\partial t} = -\frac{\partial}{\partial u^\alpha}\hat{J}^\alpha(u), \tag{A.13a}$$

crystal lattice. These are the *primitive vectors* of crystallography. T correspondence between integer points of R^2 and lattice points of E^2 complete if we take arbitrary integers for du and dv in the "different map" (A.5) to get integral linear combinations of the primitive vecto **a** and **b**.

Now, exactly as in Sect. 1.2, introduce a sufficiently smooth functio $\hat{N}(u,v,t)$ that interpolates the discrete $N_{i,j}(t)$ at integer sites $u_i = i$ an $v_j = j$. Taylor's expansion then yields

$$N_{i\pm1,j} = N_{i,j} \pm \frac{\partial \hat{N}}{\partial u}\bigg|_{i,j} + \frac{1}{2}\frac{\partial^2 \hat{N}}{\partial u^2}\bigg|_{i,j} + \ldots\ldots \tag{A.6}$$

and a similar expression in the v-direction. Insertion into Eq. (A.2) then produces the partial differential equation

$$\frac{\partial \hat{N}}{\partial t} = \frac{1}{2}\Gamma_1\frac{\partial^2 \hat{N}}{\partial u^2} + \frac{1}{2}\Gamma_2\frac{\partial^2 \hat{N}}{\partial v^2}, \tag{A.7}$$

which is "diagonal" in the oblique (u,v)-coordinates. But we are more interested in its form in physical space E^2. To find it efficiently requires a condensed notation.

Rename coordinates and basis vectors according to the scheme

$$x = x^1, \quad y = x^2, \quad u = u^1, \quad v = u^2, \tag{A.8a}$$

and

$$i = e_1, \quad j = e_2, \quad a = a_1, \quad b = a_2. \tag{A.8b}$$

Then, with Einstein's summation convention (over the range 1,2) for diagonally placed Greek indices, the equations (A.5) read simply

$$d\mathbf{x} = \mathbf{e}_\alpha dx^\alpha = \mathbf{a}_\alpha du^\alpha, \tag{A.9a}$$

where the primitive vectors are

$$\mathbf{a}_\alpha = \frac{\partial x^\beta}{\partial u^\alpha}\mathbf{e}_\beta. \tag{A.9b}$$

The four coefficients $\partial x^\beta/\partial u^\alpha$ are the components of the primitive vectors $\mathbf{a}_\alpha$ in the euclidean basis $\mathbf{e}_\beta$, and they also form the jacobian of the mapping (A.4). Indeed, these components can be computed from if the flux components (in the primitive base) are

$$\hat{J}^\alpha(u) = -D^{\alpha\beta}(u) \frac{\partial \hat{N}}{\partial u^\beta}. \tag{A.13b}$$

If we now express the interpolating function $\hat{N}$ in terms of the euclidean coordinates $(x,y) \equiv (x^1, x^2)$, then the mapping (A.4) transforms Eq. (A.12a) into

$$\frac{\partial \hat{N}}{\partial t} = A^\lambda_\alpha A^\mu_\beta D^{\alpha\beta}(u) \frac{\partial^2 \hat{N}}{\partial x^\lambda \partial x^\mu} \tag{A.14a}$$

if the A's are constants. It follows that Eqs. (A.12a) and (A.14a) have the same form if the diffusivity components in the euclidean base are

$$D^{\lambda\mu}(x) = A^\lambda_\alpha A^\mu_\beta D^{\alpha\beta}(u), \tag{A.14b}$$

which is merely the law of transformation of a second-order contravariant tensor. It is easily seen that this tensor is symmetric. Finally, similar manipulations show that Eq. (A.13a) conserves its form under the mapping (A.4) if the flux components in the euclidean base $\hat{J}^\beta(x)$ are related to those in the (u)-coordinates according to the transformation law (A.11b) for contravariant vectors. Consequently, Eq. (A.13b) is also invariant and can be written

$$\hat{J}^\alpha(x) = -D^{\alpha\beta}(x) \frac{\partial \hat{N}}{\partial x^\beta} \tag{A.14c}$$

in the euclidean base.

Exercise A.1: Connection with discrete fluxes (A.1)
The equations (A.13b) give a prescription for the computation of fluxes from the interpolating function $\hat{N}$. As in Exercise 1.1 of Sect. 1.2, show that this prescription is optimal when derivatives are evaluated at midpoints of the primitive lattice. ∎

We are now in a position to justify the diffusion equation in anisotropic media. To do so, however, requires inquiry into the physical dimensions of the quantities we have discussed. First, the numbers $N_{i,j}$ have no dimensions, and neither do the coordinates (u) of the

R^2-plane. Indeed, the area of any region in this plane is merely a measure of the number of integer lattice points that it contains. Likewise, the interpolating function $\hat{N}(u,v,t)$ has no dimensions, but, if it is to represent its discrete counterpart, then not only must it "collocate" at integer points, but also its integral over a unit cell $i-1/2 \leq u \leq i+1/2$, $j-1/2 \leq v \leq j+1/2$ (symbolized $\square$) must match. Thus,

$$N_{i,j}(t) = \int_{\square} \hat{N}(u,v,t)\,du\,dv$$

$$= \int_{ABCD} \hat{N}(x,y,t)\,|\,\partial(u,v)/\partial(x,y)\,|\,dx\,dy, \qquad (A.15a)$$

where $ABCD$ represents the interior of the contour thus named in Fig. A.1. But, according to rules of calculus, the determinant of the jacobian that figures in the second integral is the reciprocal of that formed from the elements A^{β}_{α} of Eq. (A.10), whose value is exactly the area formed by the primitive vectors. It follows that

$$\hat{N}(x,y,t)\,|\,\partial(u,v)/\partial(x,y)\,| = \hat{N}(x,y,t)/\,|\,\mathbf{a} \times \mathbf{b}\,| \qquad (A.15b)$$

is precisely what we mean by the concentration distribution $C(x,y,t)$. Therefore, Fick's first law follows if we divide Eq. (A.14c) by the constant factor $1/\,|\,\mathbf{a} \times \mathbf{b}\,|$, and the true flux (without a "hat") is given by

$$J^{\alpha}(x,y,t) = \hat{J}^{\alpha}(x,y,t)/\,|\,\mathbf{a} \times \mathbf{b}\,|. \qquad (A.15c)$$

We then obtain the continuum analogs of Eqs. (A.1-3)

$$J^{\alpha}(x) = -D^{\alpha\beta}(x)\frac{\partial C}{\partial x^{\beta}}, \qquad (A.16a)$$

$$\frac{\partial C}{\partial t} = D^{\alpha\beta}(x)\frac{\partial^2 C}{\partial x^{\alpha}\partial x^{\beta}}, \qquad (A.16b)$$

$$\frac{\partial C}{\partial t} = -\frac{\partial}{\partial x^{\alpha}}J^{\alpha}(x), \qquad (A.16c)$$

for all rectilinear (but generally oblique) lattices, and these expressions still hold in any number of dimensions by merely extending the Greek superscripts' range.

Exercise A.2: *More dimensions*

Carefully discuss the dimensions of the diffusivity tensor and of the various flux components. [Hint: What are the dimensions of A_α^β ?] ∎

B: THE PHASE RULE AND SOME OF ITS CONSEQUENCES

Some basic thermodynamic results are used to derive various forms of the phase rule and their consequences for small systems.

Any liquid-like 3-d phase is fully characterized by the $2+c$ fields T, p, and μ_i ($i = 1,...,c$), also called thermodynamic (intensive) variables. Moreover, in the absence of external fields and of macroscopic motions, these fields are constant over the domain $\mathscr{D}$ occupied by that phase if it is in equilibrium. Their values, however, are not all arbitrary, because the Gibbs-Duhem relation (2.14a) must hold at equilibrium. Therefore, only $1+c$ of these values can be arbitrarily assigned. The generalization of this simple fact to a collection of phases is called the *phase rule*, and its derivation is easy.

Consider φ contiguous 3-d phases, and assume, first, that they are all large in extent. This collection is characterized by the $\varphi(2+c)$ fields

$$T^{(\alpha)}, p^{(\alpha)}, \text{ and } \mu_i^{(\alpha)} \quad (i = 1,...,c, \quad \alpha = 1,...,\varphi), \tag{B.1}$$

which, at equilibrium, are constant over their respective domains $\mathscr{D}^{(\alpha)}$. There are now φ Gibbs-Duhem relations, one for each phase, and equilibrium conditions at the $\varphi-1$ pairs of phase boundaries

$$T^{(\alpha)} = T^{(\beta)}, \ p^{(\alpha)} = p^{(\beta)}, \ \mu_i^{(\alpha)} = \mu_i^{(\beta)} \quad (\alpha \neq \beta, \ i = 1,...,c), \tag{B.2}$$

for a total of $\varphi+(\varphi-1)(2+c)$ relations. Therefore, the number of thermodynamic variables whose values can be arbitrarily prescribed is

$$f = c + 2 - \varphi. \tag{B.3}$$

By analogy with mechanical systems, this number f is called the thermodynamic system's number of *degrees of freedom*, and Eq. (B.3) is called the phase rule. This apparently trivial result will perhaps seem less so if we realize that, having experimentally chosen the values of f variables from the list (B.1), the remaining values, as well as the values of *all other* thermodynamic variables that depend on that list (such as the entropy densities and concentrations), are entirely determined.

The phase rule's admittedly simple derivation glosses over several important issues that can cause confusion. First, it must be remembered that it holds for phases that are in global equilibrium. It *is* often used, however, in regions that are only in local equilibrium, as in the vicinity of phase boundaries. This is, in fact, an assumption on local kinetics. Next, the c components are assumed to be independent. If not (as with chemical reactions), the number c in Eq. (B.3) must be reduced by the number of linearly independent reactions. Further, it is assumed that the list of variables (B.1) fully describes the phases' internal state. This, for instance, may not be true for solids under anisotropic states of stress. Last, the phases are assumed to be large in extent, which means that capillarity effects are negligible. We now seek extensions of Eq. (B.3) when the phases are "small."

The general problem of surface effects is quite complex, because we must learn to characterize 2-d surface phases that form the boundaries of the bulk 3-d phases. This forms the subject matter of surface thermodynamics [3-6], which we broach only through the notion of surface tension γ. In other words, save for γ, the phase boundaries are assumed to have no intrinsic physical properties such as adsorbed quantities, surface energies and entropy, *etc.* Therefore, almost all the examples in this book deal with physical situations in which the boundaries are not explicitly the seat of thermodynamic changes, the only notable exception being the BCF theory of crystal growth treated in Sect. 2.6. Enumeration of surface phases is another difficulty with the general theory, because they do not exist in splendid isolation from bulk phases. Since each of these (except for pathological examples like Möbius strips) must be in contact with at least two bulk phases and since the shape of all phases is incidental, it follows that the problem of enumeration is topological in nature.

Because a system of *two* bulk phases in contact through their phase boundary is our main concern, we restrict our discussion to the case $\varphi = 2$ and to situations where γ entirely specifies the surface phase. Blithe application of the phase rule (B.3) would state simply that the number of degrees of freedom is equal to the number of independent components:

$$f = c. \tag{B.4}$$

But is this quite correct? Evidently not, because the second of the relations (B.2) must be replaced by Laplace's law (2.15b), and because we must enumerate all thermodynamic variables properly. (The next remarks are best read together with Sect. 2.4.) There are now $2(2+c)+1$ variables, because the mean curvature K must be appended to the list (B.1) to give the new list

$$T^{(\alpha)}, p^{(\alpha)}, \mu_i^{(\alpha)}, \text{ and } K \quad (i = 1,...,c, \quad \alpha = 1,2). \tag{B.5}$$

The equilibrium relations [two Gibbs-Duhem (2.14a) and the $2+c$ relations (2.15), for a total of $4+c$ relations] then lead to the special phase rule

$$f = c + 1, \tag{B.6}$$

which in fact holds true even when processes other than surface tension γ are active at the boundary of a two-phase system.

A few words are in order regarding the mean curvature's role as a thermodynamic variable. Note, first, that constant pressures $p^{(1)}$ and $p^{(2)}$ in the adjacent equilibrium bulk phases (over the domains $\mathscr{D}^{(1)}$ and $\mathscr{D}^{(2)}$, respectively; see Fig. 2.4), together with Laplace's law (2.15b), imply that K must be constant over the phase boundary Σ. Therefore, the surface's *shape* is characterized by a constant mean curvature. The study of such shapes is known as the "problem of Plateau"; it has been the object of intensive activity over the years [7,8]. Briefly, one shows that the sphere is the only connected and closed (therefore bounded) surface in 3-d space of constant and positive mean curvature. Open unbounded surfaces include the cylinder, with constant positive K, and the catenoid of revolution, with identically vanishing K. The general problem of open and bounded surfaces is both hard and interesting, as evidenced by the rich variety of soap bubbles [9]. From our point of view, namely kinetics, we must recognize that K is a field that varies over the phase boundary Σ if the adjacent phases are *not* in global equilibrium. In other words, the phase rules (B.4,6) apply to the small hatched domain in Fig. 2.4 if it is locally in equilibrium, no matter what kinetic processes occur in the core of the two adjacent bulk domains. The first (B.4) holds locally when the surface element's curvature vanishes, while the second (B.6) characterizes curved elements.

We now apply this apparatus to the estimation of a particular Gibbs-Thomson effect, namely, the influence of curvature on the melting or sublimation temperature of a pure solid. We now have $c = 1$ and, of course, two bulk phases (solid and liquid, or solid and vapor). Therefore, the phase rule (B.6) indicates that only two thermodynamic variables can be chosen at will from the seven-member list (B.5). As noted above, having chosen the values of these two variables, all others of that list, as well as any others (such as the entropy densities $s^{(\alpha)}$) whose values depend explicitly on that list, are fully determined. We choose a system such that the pressure $p^{(2)}$ of the second phase is maintained constant by contact with a "work reservoir" $p^{(e)}$. We then inquire into the functional dependence of the other variables — in particular, the melting or sublimation temperature — on the mean curvature K. The calculation is formally simple, but it is fraught with logical shoals. Let us proceed with care as in Sect. 2.4. First, we have the two Gibbs-Duhem relations (2.14a), written here for pure systems

$$s^{(\alpha)} dT^{(\alpha)} - \Omega^{(\alpha)} dp^{(\alpha)} + d\mu^{(\alpha)} = 0 \qquad (\alpha = 1,2), \tag{B.7}$$

where the atomic (or molar) volumes $\Omega^{(\alpha)}$ are the reciprocals of the total concentrations $C^{(\alpha)}$. Then, the equilibrium relation (2.15a) and the constancy of the T-fields demand the equality of the temperatures in both phases: No matter what their functional dependence on other variables might be, their values must be equal at equilibrium. Call this common value T; by definition, it *is* the equilibrium temperature of the melting or sublimation process. The same remark holds, in view of Eq. (2.15c), for the chemical potentials. Laplace's law (2.15b), on the other hand, now reads simply $dp^{(1)} = d(\gamma K)$, because we have $dp^{(2)} = dp^{(e)} = 0$, regardless of the reasons for these variations. Subtracting the two relations (B.7) from one another, and taking the previous remarks into account, we get

$$(s^{(1)} - s^{(2)}) dT + \Omega^{(1)} d(\gamma K) = 0. \tag{B.8a}$$

Now, the above difference of entropy densities is, by definition, the latent heat density L of the phase change divided by T. The previous equation then becomes a differential expression that is trivially integrated from the state of zero curvature to one of finite K ,

$$T(K) = T(0) e^{-\Omega^{(1)}\gamma K/L},$$

(B.8b)

if both L and $\Omega^{(1)}$ are independent of K. On the other hand, note that γ can have *any* curvature dependence. In contrast to Eq. (2.17c) for curvature-induced shifts of the equilibrium concentration, the melting or sublimation point *decreases* over surface elements of positive mean curvature, *i.e.*, over convex elements. The relation (B.8b), one of many regarding phase changes, is attributed to J.J. Thomson, who was not Lord Kelvin.

Exercise B.1: *One of W. Thomson's results*

Consider a one-component, two-phase (liquid-gas or solid-gas) system at constant temperature. Discuss the phase rule (B.6), and compute the effect of curvature on the vapor pressure above a curved solid or liquid element. [Hint: Assume that the vapor phase is a perfect gas.] ∎

C: MOMENTS OF DISTRIBUTIONS AND ASYMPTOTIC BEHAVIOR

In several instances we have seen that mass balance leads to the diffusion equation at interior points of a mass-containing domain $\mathcal{D}$ and to boundary conditions on internal or external boundaries of $\mathcal{D}$. The total mass contained in $\mathcal{D}$,

$$M(t) = \int_{\mathcal{D}} C(\mathbf{x},t)\,dV,$$

(C.1)

is a prototype of particular quantities called *moments*. Exactly as in probability theory [10], we shall see that the moments of the distribution $C(\mathbf{x},t)$ characterize it entirely, including its asymptotic behavior.

Let us begin, however, by proving a general result concerning fluxes. We wish to know how $\mathbf{J}$ decays at large distances when $\mathcal{D}$ is unbounded. That question is easy to answer if we assume rotational symmetry far enough from the initial distribution. If R_0 is some large enough radius, then the change in mass exterior to R_0 can only be due to the flux J_0 on the surface $r = R_0$ that is received by the system. Because one shows [11] (see also Exercise 6.12) that the area of a hypersphere in d dimensions is proportional to r^{d-1}, it follows that the volume element in Eq. (C.1) is

$$dV = \alpha r^{d-1} dr,$$ (C.2)

where α is a constant. Evaluating the time derivative of Eq. (C.1), both as the integrated flux $\alpha R_0^{d-1} J_0$ received by the system and by using the form (1.36) of the diffusion equation, we get the required asymptotic behavior

$$\lim_{r \to \infty} r^{d-1} J_r = 0.$$ (C.3)

For example, the gaussian (4.3) in 1-d,

$$G(x,t) = \frac{1}{2\sqrt{\pi Dt}} e^{-x^2/4Dt},$$ (C.4)

repeated here for convenience, evidently has this property. Indeed, not only do *all* its x-derivatives vanish for $x \to \pm\infty$, but these also vanish faster than any power of x. This is, in fact, a general property [12] of the diffusion equation's solutions: If the far field vanishes, *i.e.*, if $\lim_{x \to \pm\infty} C(x,t) = 0$, then

$$\lim_{x \to \pm\infty} x^n \frac{\partial^m C}{\partial x^m} = 0, \qquad n, m \geq 0,$$ (C.5)

which includes the property (C.3) as a special case.

Recall, now, how this gaussian serves to produce 1-d solutions of the diffusion equation for *arbitrary* initial conditions. In Chapter 4 we showed that the integral (4.5),

$$C(x,t) = \int_{-\infty}^{\infty} G(x-\xi,t)\varphi(\xi)\,d\xi,$$ (C.6)

also repeated for convenience, is precisely such a representation. Indeed, it satisfies the initial and boundary conditions

$$\lim_{t \to 0+} C(x,t) = \varphi(x) \quad \text{and} \quad \lim_{x \to \pm\infty} C(x,t) = 0.$$ (C.7a,b)

Moreover, by differentiation under the integral sign, we verify that the derivatives vanish at infinity, in accordance with Eq. (C.5). The integral representation (C.6) is sometimes called *Poisson's integral*, and, as with a similarly named integral in electrostatics and potential theory (*e.g.*, see [13, pp. 343-346]), it has an interpretation, which we now explore.

For a given "observation point" x, the gaussian $G(x-\xi,t)$, as a function of ξ, peaks sharply at its maximum $\xi = x$. There, it samples the initial distribution φ, as would a filter of width $2\sqrt{Dt}$. Varying the "source point" ξ in Eq. (C.6) over the whole range $(-\infty,\infty)$, we get the additive effects of all points of φ, which is precisely the time-developed solution $C(x,t)$. But what can we say about this solution if we are far enough (in space and time) from the initial distribution? The electrostatic analog gives a clue. If φ is a given charge distribution, then Poisson's integral (C.6) over an appropriate Green's function (the reciprocal distance $1/r$) yields the potential everywhere. Moreover, at large distances from φ, the potential appears *as if* caused by an equivalent point charge at the origin. "Equivalence" means, here, that the fictitious point charge and the real distributed charge density φ must have the same total charge, or "zeroth moment." Generally, one shows ([13, pp. 416-419]) that the potential admits a "multipole expansion" in powers of $1/r$, whose coefficients, called moments, depend only on the given charge distribution. We now develop these considerations for diffusion problems.

We first define the moments of order n

$$M_n(t) = \int_{-\infty}^{\infty} x^n C(x,t)\,dx \qquad \forall n \geq 0, \tag{C.8}$$

of the concentration distribution C. (Readers familiar with probability theory [10] will recognize that these moments are neither normalized nor centered.) Since C is some solution of the diffusion equation, we might anticipate that its moments also obey certain relations. This is indeed the case, and the proof is very simple: Differentiate Eqs. (C.8) with respect to time, and, as we have done many times, insert the diffusion equation (1.14) under the integral sign. Integrate twice by parts and use the property (C.5) to get the recursion

$$\dot{M}_n = n(n-1)DM_{n-2} \qquad \text{for } n \geq 2. \tag{C.9a}$$

Similar calculations for $n = 0$ and 1 yield simply

$$\dot{M}_0 = \dot{M}_1 = 0. \tag{C.9b}$$

Therefore, not only do we have differential equations for the moments, but these can also be integrated to give

$$M_0(t_2) = M_0(t_1) \equiv M_0, \tag{C.10a}$$

$$M_1(t_2) = M_1(t_1) \equiv M_1, \tag{C.10b}$$

$$M_2(t_2) - M_2(t_1) = 2M_0 D(t_2 - t_1), \tag{C.10c}$$

where t_1 and t_2 are two arbitrary instants of time. In other words, the zeroth and first moments are constants "of the motion," and the second moment is linear in Dt. The recursion (C.9a) can evidently be carried forward, and it is clear that the time difference of moments, of order $2n$ and $2n+1$, is an nth-degree polynomial in $D(t_2 - t_1)$. This property has been put to advantage for easy data fitting of diffusion experiments [14].

Exercise C.1: *Conditions for the moment equations*
Carefully carry through the derivation of Eqs. (C.9). What happens if C does not vanish at infinity? Interpret Eqs. (C.10) physically. ∎

Equations (C.10) are independent of the Poisson representation (C.6). In particular, however, they show that moments at time t are related to moments of the initial distribution φ. Consequently, we might expect that the full distribution $C(x,t)$ is expressible as a function of these initial moments. It is easy to find this expansion from Eqs. (C.4,6). For brevity, put $z = (x - \xi)/2\sqrt{Dt}$, and expand the exponential $\exp(-z^2)$ in a Taylor series around the point $z_0 = x/2\sqrt{Dt}$. With the known property [15]

$$\frac{d^n}{dz^n} e^{-z^2} = (-1)^n H_n(z) e^{-z^2} \tag{C.11a}$$

of Hermite polynomials ($H_0 = 1$, $H_1 = 2z$, etc.), we have

$$e^{-z^2} = e^{-z_0^2} \sum_{n=0}^{\infty} (-1)^n \frac{(z - z_0)^n}{n!} H_n(z_0). \tag{C.11b}$$

Inserting this into Eqs. (C.4) and (C.6), and remembering the meaning of the abbreviations z and z_0, we easily get

$$C(x,t) = \frac{e^{-x^2/4Dt}}{2\sqrt{\pi Dt}} \sum_{n=0}^{\infty} \frac{H_n(x/2\sqrt{Dt})}{n!\,(4Dt)^{n/2}} M_n, \tag{C.12}$$

where $M_n \equiv M_n(0)$ are the moments (C.8) of the *initial* distribution φ. Thus, as promised, the initial moments entirely characterize the concentration distribution at time t. Indeed, for large t's and moderate x's, the formula (C.12) reduces to its zeroth term, which is nothing but the gaussian (C.4) multiplied by the total (constant) mass in the system, M_0.

Exercise C.2: *Moments of gaussians*

The gaussian $(Q/\sigma\sqrt{\pi})\exp[-(x-m)^2/\sigma^2]$ depends on the three constants Q, m, σ (*cf.* Exercise 4.2). Find its moments in terms of these constants, and interpret. Is there a relation between moments? Why? ∎

The message of the expansion (C.12) should be clear: For long enough times in an infinite medium, any initial distribution will decay into a gaussian, proportional to Eq. (C.4), with corrections proportional to $1/t^{n/2}$. Moreover, for the analogous electrostatic problem, it is recalled that an appropriate choice (at the center of charge) of the origin of coordinates cancels the "dipole" term. Likewise, here we may hope that there exists an origin of space and time, and a scale factor for C, such that some leading terms drop out of the moment expansion (C.12). This should be possible because our problem is translationally invariant in space and time, and it is also homogeneous (*cf.* Properties C of Sect. 1.3). Physically, if we can find a leading gaussian that mimics the initial distribution (and the last exercise showed that there are precisely three parameters to choose from), then we expect, at least, "osculating" behavior of the series (C.12).

Assume, therefore, that C develops from some *fictitious* initial condition ψ that holds at some earlier time $t = -t_0$:

$$C(x,t) = \frac{1}{2\sqrt{\pi D(t+t_0)}} \int_{-\infty}^{\infty} e^{-(x-\xi)^2/4D(t+t_0)} \psi(\xi)\,d\xi. \tag{C.13}$$

Consequently, the *real* initial distribution φ can also be expressed in terms of ψ by putting $t = 0$ in this Eq. (C.13):

$$\varphi(x) = \frac{1}{2\sqrt{\pi D t_0}} \int_{-\infty}^{\infty} e^{-(x-\xi)^2/4Dt_0} \psi(\xi) \, d\xi. \tag{C.14}$$

Now, exactly as with Eqs. (C.11), expand Eq. (C.13) around the as yet unspecified point $z_0 = (x-x_0)/[4D(t+t_0)]^{1/2}$ to get

$$C(x,t) = \frac{e^{-z_0^2}}{2\sqrt{\pi D(t+t_0)}} \sum_{n=0}^{\infty} \frac{H_n(z_0)}{n! \, [4D(t + t_0)]^{n/2}}$$

$$\times \int_{-\infty}^{\infty} (\xi - x_0)^n \psi(\xi) \, d\xi. \tag{C.15}$$

Next, define the moments

$$N_n = \int_{-\infty}^{\infty} x^n \psi(x) \, dx \tag{C.16}$$

of the fictitious initial distribution. But we have no desire to characterize fictitious distributions. Our task, therefore, must be to express the integrals in the expansion (C.15) in terms of the moments M_n of the real initial distribution φ, and, along the way, to choose the undetermined constants x_0 and t_0 appropriately.

This is easily done because Eq. (C.14) connects φ to ψ. Their moments must likewise be related, in the spirit of Eqs. (C.10), and, setting $t_2 = 0$, $t_1 = -t_0$, we have

$$M_0 = N_0, \quad M_1 = N_1, \quad M_2 = N_2 + 2M_0 D t_0, \quad etc. \tag{C.17}$$

Let us calculate the leading integrals of the expansion (C.15). For $n = 0$ we have

$$\int_{-\infty}^{\infty} (\xi - x_0)^0 \psi(\xi) \, d\xi = N_0, \tag{C.18a}$$

which, by the first equation (C.17), equals M_0. Therefore, the expansions (C.12) and (C.15) have the same total mass, a not altogether unexpected result. Next, for $n = 1$, we have

$$\int_{-\infty}^{\infty} (\xi - x_0)^1 \psi(\xi) \, d\xi = N_1 - x_0 N_0 = M_1 - x_0 M_0 \tag{C.18b}$$

by the same arguments. This term vanishes if we choose $x_0 = M_1/M_0$, *i.e.*, the origin must be at the center of gravity of the real initial distribution, as expected from the electrostatic analog. Finally, for $n = 2$, we have

$$\int_{-\infty}^{\infty} (\xi-x_0)^2 \psi(\xi)\,d\xi = N_2 - x_0^2 N_0$$

$$= M_2 - 2M_0 Dt_0 - x_0^2 M_0, \qquad (C.18c)$$

and this "moment of inertia" vanishes for a proper choice of t_0. In sum, in an infinite medium, the gaussian

$$\frac{Q}{2\sqrt{\pi D(t+t_0)}} \exp\left[-\frac{(x-x_0)^2}{4D(t+t_0)} \right] \qquad (C.19a)$$

is an asymptotic representation of the solution C, up to terms of order $[4D(t+t_0)]^{-3/2}$, if

$$Q = M_0, \quad x_0 = M_1/Q, \quad t_0 = \left(M_2 - M_1^2/Q\right)/2DQ, \qquad (C.19b)$$

and there are no other free parameters at our disposal. The choice (C.19b), in fact, forces the gaussian (C.19a) to match the real initial distribution φ through the equality of their first three *central* moments. This gaussian itself is the result of a fictitious *point* source ψ, of strength Q, placed at the point $x = x_0$ when $t = -t_0$.

REFERENCES

1. A.J. McConnell, *Applications of Tensor Analysis* (reprinted by Dover, New York 1957).
2. R. Aris, *Vectors, Tensors, and the Basic Equations of Fluid Mechanics*, especially Chap. 7 (Prentice-Hall, Englewood Cliffs, NJ 1962).
3. J.W. Gibbs, "On the Equilibrium of Heterogeneous Substances," *Trans. Conn. Acad.* (1875-1878). [Reprinted in *Scientific Papers*, Vol. 1 (Dover, New York 1961).]
4. R. Defay and I. Prigogine, *Tension Superficielle et Adsorption* (Editions Desoer, Liège 1951). [Engl. transl. by D.H. Everett, John Wiley, New York 1966].
5. F.P. Buff, "The Theory of Capillarity," in *Handbuch der Physik*, Vol. 10, pp. 281-304 (Springer-Verlag, Berlin 1960).

6. R. Ghez, "A Theoretical Crystal Grower's View of Phase Equilibria," in *Int'l School of Crystallography, Erice, 1987* (NATO A.S.I.), to be published (1988).

7. R. Osserman, *A Survey of Minimal Surfaces* (reprinted by Dover, New York 1986).

8. W.H. Meeks III, "A Survey of the Geometric Results in the Classical Theory of Minimal Surfaces," *Bol. Soc. Bras. Mat.* **12**, 29 (1981).

9. C.V. Boys, *Soap Bubbles, Their Colours and the Forces Which Mold Them* (reprinted by Dover, New York 1958).

10. B. Gnedenko, *The Theory of Probability*, pp. 185-191 (Mir Publishers, Moscow 1976).

11. C. Kittel, *Elementary Statistical Physics*, p. 37 (John Wiley, New York 1958).

12. E. Goursat, *Cours d'Analyse Mathématique*, Tome III, pp. 287-322 (Gauthier-Villars, Paris 1956).

13. A.N. Tikhonov and A.A. Samarskii, *Equations of Mathematical Physics* (Pergamon Press, Oxford 1963).

14. R. Ghez, J.D. Fehribach, and G.S. Oehrlein, "The Analysis of Diffusion Data by a Method of Moments," *J. Electrochem. Soc.* **132**, 2759 (1985).

15. N.N Lebedev, *Special Functions and their Applications*, R.A. Silverman, Ed., pp. 60-76 (reprinted by Dover, New York 1972).

INDEX

Supersaturation:
  absolute, 68, 127
  due to temperature ramping, 127
  relation to critical radius of nucleation,
    54, 59
  relation to step distance, 73, 77
  relative, 68, 77
Symmetries:
  affine, 14, 38,
  geometrical, 29
  rotational, 30
  see Invariance

Temperature schedule, 125, 204
Thermodynamic:
  equilibrium, 56, 229
  local equilibrium, 64, 230, 231
  reservoirs, 57, 232
  variables, 229
Transition frequency,
  see Jump frequency